A Photo Gallery

BUCYRUS HEAVY EQUIPMENT

Construction and Mining Machines 1880-2008

Keith Haddock

Iconografix

Iconografix
PO Box 446
Hudson, Wisconsin 54016 USA

Library of Congress Control Number: 2008930629

ISBN-13: 978-1-58388-219-1
ISBN-10: 1-58388-219-7

08 09 10 11 12 13 6 5 4 3 2 1

Printed in China

Cover and book design by Dan Perry

Copyedited by Andy Lindberg

On the cover: A Bucyrus Erie 3½ cubic yard 71-B shovel works on a highway job in Colorado.

Table of Contents

Acknowledgments

Of all the nine books I have written on heavy equipment, including two jointly with other authors, this Bucyrus book gave me the most pleasure and satisfaction during its preparation. My association with the Bucyrus companies spans many years from my first visit to the Lincoln, England factory in 1959, through my management years on earthmoving projects and surface mining operation in the U.K. and Canada, to the several more recent visits to South Milwaukee offices, as I gathered information for my two Bucyrus books. Over the years it has been my pleasure to have been associated with many fine professional and sincere individuals from the Bucyrus organization. In particular I would like to thank the following:

- Kent Henschen, Manager – Marketing, Bucyrus International, Inc. and his staff Jennifer Okonek and Ron Gratzke. Thanks, Kent, for your continued help and support for my endeavors.
- David M. Lang, Director of Engineering – Shovels and Drills, Bucyrus International, Inc. Your help in providing engineering details and your personal interest in Bucyrus history have been most valuable.
- Peter A. Cottam, Sales Manager, Bucyrus Europe Ltd. Thanks for keeping me in touch with Bucyrus' activities in the other half of the world.
- Derek Broughton, Museum of Lincolnshire Life and former employee of Ruston-Bucyrus Ltd. Derek deserves a medal for his tireless work in preserving the R-B Lincoln archives throughout the drastic corporate changes of the past decade.
- Eric C. Orlemann, ECO Industrial Communications, author and friend. Thanks again for your expert advice and encouragement which has continued since my first book.
- Last but not least, I give sincere thanks to my wife Barbara for her support, advice and patience during the hours each day I work in my home office to produce these publications. Barb also edits all my work before it goes to the publisher.

Foreword

I feel most privileged to have this opportunity to document and write the fascinating history of one of the most amazing companies in the construction equipment world.

The present company, Bucyrus International, Inc., of South Milwaukee, Wisconsin was known as Bucyrus-Erie Company (B-E) from 1927 to 1996, a familiar name I am sure to the majority of readers of this book. Before 1927 the company operated under other corporate names and was originally established as the Bucyrus Foundry and Manufacturing Company in 1880. Bucyrus has enjoyed a long and colorful history of excavating machine manufacture reaching back to the start of industrial mechanization. No other company built such a wide variety of excavating and lifting machines, and few can boast a continuous manufacturing history of 128 years.

My first book about this company "Bucyrus – Making the Earth Move for 125 Years" was a corporate history book to commemorate that milestone, highlighting achievements over the decades. This latest book focuses on the machines themselves. It covers every model of excavator and most of the vast ranges and types of other machines built by the company, from the smallest loader-backhoe to some of the largest machines ever to move on land. Because the corporate management and machines built by the joint venture Ruston-Bucyrus Ltd. (R-B) at Lincoln, England, are so closely intertwined with the American company, the R-B machines are also included. In many cases R-B models outsold their American B-E counterparts. The Ruston machines prior to 1930 and the R-B machines after 1985, with one or two exceptions, are excluded as they are not strictly Bucyrus machines.

This book is intended to be an illustrated reference source for everything Bucyrus. For quick reference the machine descriptions are arranged in ascending size, and each chapter covers a specific machine category. As you read the machine descriptions with each photograph, you may also refer to the appendices at the back to locate production numbers and manufacturing dates along with other information.

I hope you enjoy reading about the machines from one of America's oldest but innovative, diverse yet specialized, industrial companies. I am excited to present to you this book about the machines of the Bucyrus companies.

KOMATSU-BUCYRUS

BUCYRUS-ERIE

RUSTON-BUCYRUS

BUCYRUS-ERIE COMPANY

BUCYRUS®

Chapter 1
Corporate History
The Early Years

The Bucyrus Foundry and Manufacturing Company was established in December 1880 by a group of prominent business associates led by entrepreneur Daniel P. Eells of Cleveland, Ohio. The group purchased the former Bucyrus Machine Company at Bucyrus, Ohio, and set out to serve the railroad and mining industries. In 1882 the company received its first steam shovel order from the Ohio Central Railroad. The machine was known as the "Thompson Iron Steam Shovel and Derrick," named after its designer John Thompson, Manager of Manufacturing at Bucyrus.

The success of Thompson's shovel, sold in the heyday of railroad construction, generated rapid sales for the company which was able to expand into manufacturing other types of excavating machinery. The vast array of products included railroad wrecking cars, ballast unloaders and spreaders, steam pile drivers, mine cars and dredges. In 1883 Bucyrus shipped its first dipper dredge, and for seven decades dredges played a significant part in the company's activities.

Move to South Milwaukee

By the early 1890s the production facilities at Bucyrus, Ohio, proved to be increasingly inadequate as the product line was shifting toward larger machines. As additional land adjacent to the existing Ohio site could not be purchased, a new location was sought. An attractive site, perfectly fitting the company's needs, was found in South Milwaukee, Wisconsin. Plans were immediately put into action to uproot the established company in Ohio and move it to the new location.

President Ted Roosevelt boards a Bucyrus shovel during a 1908 inspection tour of the Panama Canal construction.

Seventy-seven Bucyrus Shovels worked on the Panama Canal including 32 of the giant 95-ton size, largest available at the time.

Operations at the new South Milwaukee plant began in April 1893, and in August that same year the company changed its name to Bucyrus Steam Shovel and Dredge Company of Wisconsin. The first few years at the new plant in South Milwaukee proved very difficult because of problems training the new unskilled labor force and a lower demand for the company's products. Financial difficulties led to reorganization of the company in 1896 and another name change to The Bucyrus Company.

Despite these difficulties, Bucyrus shovels and dredges earned a solid reputation. From 1895 through 1901, a total of twelve different sizes of steam shovels were offered, ranging from a 12-ton model to a gigantic 95-ton machine for use in mining and heavy railroad excavation. All of these were the partial swing "Railroad" type of shovel mounted on standard-gauge railroad tracks.

Major Construction Projects

Bucyrus' reputation continued to grow as the company participated in major excavation projects across the United States and around the globe. Twenty-four Bucyrus shovels were used to dig the Chicago Drainage Canal and more were shipped to the vast open-pit Mesabi Iron Ranges. Bucyrus dredges were shipped to the California gold fields and for enlargement of the New York State Barge Canal, the largest earth-moving project in the U.S. between 1902 and 1912. But the most prestigious

Bucyrus purchased competitor Vulcan Steam Shovel Company of Toledo, Ohio, in 1910. Vulcan had already shipped fully-revolving shovels like this one, ahead of most of its competitors.

project in this period was the Panama Canal. When America took over the Canal in 1904, no time was wasted in getting the excavation work restarted after the failure of the earlier French attempt. By the time the last shovel was shipped to Panama in 1908, Bucyrus had delivered 77 out of the 102 steam shovels on the project.

First Draglines

Draglines came to the fore in the early part of the 20th century. They were born in the Chicago area where extensive drainage canals were being dug. Bucyrus observed this development with great interest and recognized the importance of these machines. In 1910 Bucyrus purchased the manufacturing rights of the Heyworth-Newman Dragline Excavator from drainage contractor James O. Heyworth who had been building and using his own machines in the Chicago area.

The following year, Bucyrus introduced the Class 24 as the world's largest dragline carrying a 3 1/2-cubic yard bucket on a 100-foot boom. These early draglines were mounted on rails or skids and rollers—walking draglines were still a few years away.

A full range of similar models soon followed including, in 1912, a Class 14, the first Bucyrus excavator to be equipped with crawler tracks.

Expanding Horizons

Looking to expand its manufacturing base beyond a single plant, The Bucyrus Company established a manufacturing affiliate with Poutilov Works of St. Petersburg, Russia, in 1900, and one with the Canada Foundry Company of Toronto in 1904. Although several machines were built by these companies, both ventures were only marginally successful and were all but discontinued by 1910. Difficulty in manufacturing satisfactory machines with the quality and reliability of their American-built counterparts was the stated reason for their failure.

In 1910, Bucyrus acquired its major competitor, the Vulcan Steam Shovel Company of Toledo, Ohio, and established the Bucyrus-Vulcan Company. Vulcan's predecessors were founded in 1872 and its steam shovels ranged from 40 to 120 tons in size including some fully revolving types. The deal not only gave Bucyrus access to Vulcan's steam shovel line, but also a new 70-acre site at Evansville, Indiana, already in negotiation by Vulcan just before the merger. Bucyrus followed through with the property arrangement, and lost no time in constructing an extensive new plant. The Evansville plant was designated to build "construction-size" excavators, and its first products (5/8-yard 14-B and 7/8-yard 18-B) were ready for testing as early as the spring of 1912.

Only a year after establishment of Bucyrus-Vulcan, another competitor was consolidated into the group. This time it was the Atlantic Equipment Company, a builder of steam shovels since 1902. It had already gained a fine reputation for building heavy-duty railroad-type steam shovels designed by A.W. Robinson, former chief engineer at Bucyrus. With the merger of Bucyrus, Vulcan and Atlantic, the new corporate name became simply "Bucyrus Company."

Bucyrus During World War I

Bucyrus took on a major contract for the British Government to supply 8-inch shells. The new Evansville plant was retooled to make these, and 58,657 shells were produced in 1916 and 1917. A portion of the British shell order and contracts for other special equipment were manufactured at the South Milwaukee plant. The special equipment and experience gained from the British contracts were useful in obtaining substantial orders for the American war effort. Equipment supplied to the United States Government up to 1919 included forty-seven 15-ton (14 tonne) locomotive cranes, twenty self-propelled derricks, and large numbers of special forgings and gun components for the U.S. and French armies.

Post World War I to 1927

In 1920, Bucyrus introduced its first "universal" excavator, the 1-cubic yard crawler-mounted 30-B. "Universal" is the industry term for machines that could be converted from the basic shovel to dragline, clamshell or crane with minimal alteration to the machinery. The 30-B was initially a steam shovel but a diesel engine was offered in 1921, making this the first diesel-driven shovel in the industry. That same year, the similarly-designed 20-B and larger 50-B shovels were unveiled, making a trio of very successful excavators for the company, and paving the way for even more popular machines to come.

Meanwhile at the main South Milwaukee plant, rapid progress was being made with several large surface mining machines. There is no doubt that the introduction of stripping shovels and draglines enabled rapid expansion of surface mining, proving it to be the most efficient and low-cost method of mining in certain geological conditions. Between 1914 and 1927, the tonnage of coal produced by such methods expanded fourteen fold, while the output from underground mines rose only 20%. Bucyrus delivered its first stripping shovel, a 150-B with a 2 1/2-cubic yard dipper in 1912, and in 1914 introduced the 6-yard 225-B of which 90 units had been sold by 1923. It was evident that the large, fully revolving shovel and dragline had created an entire new industry. By 1927, sizes of these machines had increased five fold since inception and, as we now know, future machines would reach a size beyond the wildest imagination of engineers at that time.

In 1925, Bucyrus introduced a significant new class of machine, the 5-yard 120-B quarry and mine shovel, first to combine the advantages of a fully revolving shovel with the proven heavy-duty digging capability of a part-swing railroad shovel. This machine was the forerunner of the popular present-day electric mining shovels.

The Merger with Erie

In 1927, Bucyrus merged with the Erie Steam Shovel Company of Erie, Pennsylvania, and the new corporate identity of Bucyrus-Erie Company resulted. The 1920s was the decade of the small universal revolving excavator, and this merger was the culmination of efforts to compete with dozens of new manufacturers of these machines. The Erie

Steam Shovel Company began as the Ball Engine Company in 1883 as a manufacturer of steam engines and built its first steam shovel in 1914. With its concentration on small machines and its streamlined efficient manufacturing, Erie was America's leader in small excavators. The merger gave Bucyrus-Erie Company a strong presence in the small machine market, something it previously lacked, and an additional major manufacturing plant at Erie, Pennsylvania, which would be devoted to small "construction size" excavators.

The Merger with Ruston

In 1930, Ruston-Bucyrus Ltd. (R-B) was established at Lincoln, England, with ownership split almost equally between Bucyrus-Erie Company (B-E) and the British firm Ruston & Hornsby Ltd. who, with its predecessors, had built steam shovels since 1874. When the merger occurred, Ruston & Hornsby was the only company outside the United States offering a full line of excavators of all sizes up to the giant stripping shovels. Earlier products had included steam traction engines, pumps, oil-field equipment, steamrollers and locomotives. Over 70 Ruston steam shovels had been employed on the Manchester Ship Canal in England, beginning in 1887, the first ever project to use powered excavators in considerable numbers.

Within a few years the R-B management consolidated the lines of both companies, but selected only certain models from the Bucyrus-Erie line to be built at Lincoln. Those models selected for U.K. manufacture, however, often sold in far greater numbers than their U.S.-built counterparts. For example the 10-RB sold 7,625 units vs. 2,428 for the 10-B, the 33-RB sold 527 units vs. 96 for the 33-B, and the 22-RB sold 10,245 units vs. 8,828 for the 22-B. The merger also enabled B-E to expand internationally with R-B acting as sales agents for those B-E models not made at Lincoln, opening markets in countries where Bucyrus had previously been inactive.

Walking Draglines

The non-walking draglines Bucyrus marketed in 1910 did the job they were designed for. In fact hundreds were sold over the next two decades including in the 1920s some giant machines weighing up to 500 tons. Mounted on skids and rollers, crawlers, or on rail trucks, they were obviously restricted in mobility. A vastly improved system of moving bulky draglines was invented in 1913 by Oscar Martinson, chief engineer of the Monighan Manufacturing Corporation of Chicago, Illinois, a specialist dragline builder. The ingenious device employed a pair of large shoes, one on each side of the machine, moving in unison to raise and propel the machine in a walking motion. Well aware of its obvious advantages—freedom from rails or restrictions, lower ground pressure, and ability to walk in any direction, B-E began purchasing shares in the company and gained full control of the Monighan Manufacturing Corporation in 1934. The Bucyrus-Monighan Company was thus established with Oscar Martinson as president.

The Bucyrus-Monighan Company with its Chicago plant functioned as an independent unit until 1949 when Martinson retired as president and B-E took overall control of the company. The plant finally closed in 1958 and walking dragline manufacture transferred to South Milwaukee. Many of the Monighan design features were carried forward into new larger walking draglines of the 1950s and 1960s, and the same design philosophy can still be found in the giant walkers shipped today by Bucyrus International, Inc.

Armstrong Drill Company

B-E purchased the Armstrong Drill Company of Waterloo, Iowa, in 1933. Armstrong could trace its manufacturing roots back to 1868, and offered a line of oil-well, water-well and blast-hole mobile drills. These were well suited to B-E's existing customers who were often involved with rock excavation that required blasting before it could be excavated. The drills were sold as Bucyrus-Armstrong products up to 1943, and thereafter as Bucyrus-Erie products. Today, blast-hole drills are one of Bucyrus International's three key products to serve the surface mining industry, along with shovels and walking draglines.

Tractor Equipment

Bucyrus-Erie designed and produced tractor equipment (scrapers, bulldozer blades and winches) beginning in 1935. This was intended to capture a share of the market made possible by the larger diesel crawler tractors being developed

Railroad cranes formed an important part of Bucyrus' machine production in the first half of the 20th century. This steam-powered 120-ton is a typical example.

in the 1930s, and which in some cases were stealing potential excavator sales from B-E. Thousands of tractor attachments were shipped worldwide for use during and after World War II. They were mostly fitted to crawler tractors made by International Harvester Company with whom B-E formed an alliance. Later on, the main-line tractor makers started building their own tractor equipment, and in 1954 B-E withdrew from this market which had been very profitable for the company.

Bucyrus-Erie in World War II

With the outbreak of World War II, Bucyrus-Erie experienced an increased demand for its excavators. The Company fortunately did not have to re-tool for new products to any great extent as its standard products were what the Government needed. But it did produce some special ordinance equipment including gun carriages. In the war years 1942 to 1945, some two-thirds of the Company's production was devoted to the war effort. Its small excavators sold in large quantities, the 1/2-cubic yard 15-B universal excavator being the most popular with the armed forces.

Hydraulic Machines

B-E purchased the Milwaukee Hydraulics Corporation in 1948 along with its H-2 truck-mounted hydraulic crane with a telescopic boom and hydraulically operated outriggers. Invented two years earlier by Roy O. Billings, president of the company, it was one of America's first-ever hydraulic cranes. It was Bucyrus-Erie's first move into hydraulic machines and paved the way for a full line of hydraulic excavators and truck cranes that would sell in large numbers in the 1960s and 1970s. The

Six Monighan walking draglines dig a drainage ditch in Mississippi in 1934. Monighan draglines were added to the Bucyrus roster in the early 1930s.

truck-mounted Hydrocrane product, and backhoe version, the Hydrohoe, were upgraded over the years and remained in Bucyrus-Erie's product line until 1981.

Another company building hydraulic machines, the Hy-Dynamic Company of Lake Bluff, Illinois, was acquired by B-E in 1971. This company produced the Dynahoe, a large version of the popular tractor-loader-backhoe combination machine. The Dynahoe became a B-E product and the concept was expanded into several machine sizes which remained in the line until 1985.

Dredges

Dredges formed a significant part of B-E's business in the early years. The period 1889 to 1910 was the most prolific for dredge manufacture when the Company shipped an average of twelve dredges every year. After that time, dredge shipments leveled out to three and four per year up to World War II. After the war, dredge orders dwindled and only five more were sold up to 1950. The longevity of the machines and a decrease in gold dredging, caused market saturation, and B-E officially withdrew from the market in 1953, selling the dredge business in 1957.

Bucket Wheel Excavators

Bucyrus-Erie built some large bucket wheel excavators during its fascinating history. These machines offered maximum output due to their continuous digging capability, i.e., they were not dependent on the usual intermittent cycle of dig-swing-dump-return. The bucket wheels built by B-E were known as the "cross-pit" type because they excavated material from one side of the "cut" and deposited it on the opposite side of the cut in one smooth motion. B-E installed several of these

The 2 1/4-yard 75-B was the smallest of the quarry and mine shovels in the early 1930s. It could also be converted for dragline duty.

lumbering monsters the 1950s and 1960s, and then in 1986 built the world's largest of this type, the 5,380-ton monster at the Captain Mine, Illinois.

Giant Machines of the 1950s and 1960s

Perhaps B-E's most spectacular machines are the super-sized stripping shovels. B-E sold its first "Super Stripper," the 55-cubic yard 1650-B stripping shovel known as the "River Queen," to Peabody Coal Company, who put it to work in 1956. The boom in coal-fired electric power generation was underway and many more super-sized machines were put in service over the subsequent decade. The largest shovels B-E ever built were two model 3850-Bs, also sold to Peabody Coal Company. They were commissioned in 1962 and 1964 with dipper capacities of 115 and 140 cubic yards respectively.

The most famous of all B-E's machines was the 4250-W walking dragline known as "Big Muskie." It carried a whopping 220-cubic-yard bucket on a 310-foot boom, and went to work in 1969 as one of the largest mobile earthmoving machines ever built. It operated at the Muskingum mine of the Central Ohio Coal Company until 1991, and sadly was scrapped in 1999.

International Operations

The Ruston-Bucyrus joint venture in England continued to significantly contribute to worldwide sales each year since its establishment in 1930. In the 1940s Tornborg & Lundberg of Sweden took on the manufacture of Ruston-Bucyrus excavators under license in that country. In 1957 Bucyrus-Erie Company of Canada, Ltd., a wholly owned subsidiary of Bucyrus-Erie Company, commenced

The innovative 950-B stripping shovel appearing in 1935 employed a counterbalanced hoist system, where a moving counterweight balanced the weight of the empty dipper.

production of excavators at a new plant in Guelph, Ontario. Prior to this, the 22-RB excavator had been fabricated and assembled in Canada since 1949 at the Maritime Steel and Foundries plant at New Glasgow, Nova Scotia. Some 40 people were employed there to produce 361 units over the seven-year period.

Expansion overseas continued throughout the 1960s, and in its report for 1964 the Company boasted, "B-E products are manufactured by joint venture affiliates and licensees in Australia, France, India, Japan, Mexico, New Zealand, Sweden and South Africa."

In Japan the joint venture Komatsu-Bucyrus was established in 1963 to build cable cranes and excavators, with models 22-KB and 25-KB the most popular. Hydraulic excavators commenced in 1967 with the 15-HT followed by the 20-HT. In Brazil, B-E's subsidiary BECO entered into a joint venture with Fabrica Nacional de Vagoes, S.A. (FNV) to build construction excavators and cranes starting in 1965. In Mexico, Equipos Mecanicos Bucyrus-Erie, S.A. was formed in 1963 to build two excavator models for the home market.

Rapid Expansion in the 1970s

The 1970s was an extremely busy decade for Bucyrus-Erie. The Arab Oil Embargo of October 1973 caused the energy supply industry to turn to coal as the most reliable and abundant source of energy for power generation and to reduce America's dependency on foreign oil. Electric utilities moved to secure additional coal sources and oil companies scrambled to purchase coal reserves and open mines. The result was an unprecedented demand for B-E's mining shovels, walking draglines and drills.

This chaotic market situation was such that by

Large blast hole drills form a major portion of the current products of Bucyrus International, Inc. This late model 45-R could drill holes to 10 5/8 inch diameter.

the end of 1973, the volume of orders received by B-E would require more than fifteen years to deliver running its existing production capacity at an absolute maximum. With prospects of the backlog rising even further, Bucyrus-Erie management chose to accept the challenge to fulfill all its incoming orders within a reasonable time frame and formulated a plan that would double its manufacturing space in less than a year!

So during 1974, the South Milwaukee plant added machine tools and extended existing buildings, and a machine shop and foundry was purchased at Glassport, Pennsylvania. A new facility was purchased in Racine, Wisconsin, to manufacture electrical components, and this became B-E's world Central Parts Depot. A huge former Naval Ordnance plant was also acquired at Pocatello, Idaho, with 1,400,000 square feet of manufacturing space under roof. In addition, the Erie and Evansville plants were modified to build mining machine components, and B-E's overseas manufacturing partners in England, Australia, Japan and South Africa also filled orders for mining shovels and walking draglines.

Rapid Contraction in the 1980s.

The 1980s was another tumultuous decade for B-E. But whereas the 1970s turbulent activities resulted in expansion, by sharp contrast the 1980s resulted in severe contraction.

In March 1981, the Komatsu-Bucyrus joint venture active in Japan since 1963 was dissolved, and B-E sold technical data pertaining to certain hydraulic excavators to Komatsu. But later that year, a new Japanese venture was started with Mitsui Engineering & Shipbuilding Co. Ltd. for that company to build and market mining shovels. Also in 1981, a similar manufacturing agreement for mining shovels was established with Heavy Engineering Corporation (HEC) of Ranchi, India.

B-E continued to enjoy full production of its expanded facilities for about five years after 1974. But the widely expected shift of electric utility fuel from oil to coal failed to materialize, as government regulations either prevented it outright or made coal conversions uneconomical. With uncertainty surrounding long-term coal contracts, utilities and contractors began to cancel large numbers of orders.

The recession of the early 1980s was more than most earthmoving equipment manufacturers could survive. Great changes occurred across all industries, with bankruptcies, mergers and takeovers being reported almost every month. B-E only survived by reducing in size, consolidating all manufacturing at its South Milwaukee plant, and transforming itself into a company serving only the surface mining industry.

Plants were sold as quickly as possible beginning with Glassport in 1981, and Racine and Evansville the following year. The massive Pocatello plant was vacated in 1983 after only nine years of production, and that same year the South Milwaukee Foundry was closed. The Erie plant was sold in 1984. Finally, in 1985, the Company sold its share in Ruston-Bucyrus Ltd. Ruston-Bucyrus Ltd. continued to produce machines in England under new ownership and the new name of R-B Lincoln plc. All rights to Bucyrus-Erie's non-mining products, including truck cranes, rough-terrain cranes, tractor-backhoes and construction-size hydraulic and cable excavators were either sold to other companies or discontinued by 1985. The marine crane division was sold in 1997. The remaining products (electric

The stalwart 88-B was the largest of the "construction size" excavators, remaining in production from 1946 to 1984. It was convertible from shovel to dragline.

shovels, walking draglines and rotary blasthole drills) became the company's key products and are continually updated to serve the increasing needs of the surface mining industry.

In an effort to counteract the cyclic nature of the mining equipment business, Bucyrus-Erie purchased Western Gear Corporation, a manufacturer of aerospace products, in 1981. To recognize this diversity, the company changed its name to Becor Western, Inc. in 1985, but the Bucyrus-Erie Company name was retained for its mining machinery division. Two years later, Western Gear Corporation was sold, and in a leveraged buyout in 1988 the Company was purchased by B-E Holdings, Inc., a group of private investors.

Key Competitors Purchased

Bucyrus-Erie continued its acquisition of key companies that would contribute to overall success in the surface mining industry. In 1988, it acquired the dragline assets of Ransomes & Rapier Ltd. of England, affording greater participation in the Indian market where most of that company's large draglines have been sold. Established in 1869, Ransomes & Rapier built an array of heavy equipment including its first walking dragline in 1939, and the world's largest, the W1800, in 1960. From 1924 to 1936, it built excavators under license from the Marion Steam Shovel Company, a company that would become part of the Bucyrus organization in 1997. Bucyrus Europe continues to market the Rapier design walking draglines with primary sales continuing in India.

In 1997, Bucyrus-Erie purchased its main competitor, Marion Power Shovel Company, after 113 years of rivalry. Marion had been established in 1880 as the Marion Steam Shovel & Dredge Company and grew into one of the foremost manufacturers of excavating machines. Marion built the first long-boom revolving stripping shovel in North

The first series of Bucyrus' hydraulic excavators is represented here by the 15-H, 20-H and 30-H in this 1967 publicity shot.

America in 1911, and its first walking dragline in 1939. The most famous Marion machine was the behemoth 6360 stripping shovel at the Captain Mine, Illinois. This was the world's largest shovel of any type, weighing approximately 15,000 tons and utilizing a 180-yard dipper. Marion also made headlines when it built the famous Apollo moon rocket transporters for NASA in 1965, based on stripping shovel undercarriage technology.

A New Name for the 21st Century

In 1996, the name Bucyrus-Erie Company was changed to Bucyrus International, Inc. to better reflect its international status, with the highest percentage of sales being shipped outside the United States.

With 128 years of history behind it, and now able to draw on the best design features from the highly respected Marion machines, Bucyrus International serves the worldwide surface mining industry with leading-technology machines, high-quality parts and, through its subsidiary company Minserco, professional contract services to perform major rebuilds and machine moves.

As its latest acquisition, Bucyrus International, Inc. announced in early 2007 the purchase of DBT GmbH, a leading manufacturer of underground mining equipment based in Lunen, Germany. With DBT annual sales similar in volume to those of Bucyrus, the DBT acquisition is expected to further strengthen the company's ability to support the overall worldwide mining industry.

Chapter 2
Cable Excavators

The cable excavators of Bucyrus-Erie are the most important and well-known of all the company's products, and range from the earliest railroad steam shovels made by Bucyrus Foundry & Manufacturing Company, to the large electric mining shovels produced by Bucyrus International, Inc. today. This chapter covers universal excavators (convertible to shovel, dragline, backhoe, etc.) commonly mounted on crawler tracks, and machines at the lower end of the range generally classified as "construction-size" excavators (models 10-B to 88-B). It briefly describes every excavator in this category produced by the Bucyrus companies including those built by Erie Steam Shovel Company and British models built by Ruston-Bucyrus Ltd. Excluded are the Ruston line of machines before the formation of Ruston-Bucyrus in 1930, and the hydraulic cranes and excavators made by R-B (Lincoln) plc after Bucyrus-Erie's involvement.

Following Bucyrus Company's acquisition of Erie in 1927, a preponderance of machines resulted as newly designed models from both companies hit the market, while old models were kept current. By the early 1930s, the line consisted of a confusing array of different designs of small excavators 1 1/2-cubic-yards capacity or less, powered by steam, gasoline, diesel, electricity and gas+air. This diversity of products concerned company chairman W.W. Coleman whose instructions resulted in consolidation of machines in the 1930s. These eventually led to a new sleek line of models with welded construction replacing the former riveted designs.

Up to the 1960s, as with most equipment manufacturers, revised models received new model number designations. Since then, it became the fashion to retain familiar model number and identify the progressive upgrades by use of series or suffixes, e.g., 30-B Series II, 30-B Series III, etc.

Over time, shovels and draglines became larger so that by the early 1970s the larger machines, mostly used for surface mining, became the responsibility of Bucyrus-Erie's Mining Equipment Division. The smaller machines, 88-B and smaller, were manufactured and marketed under the Construction Equipment Division. This separation of products would become very significant in the 1980s when Bucyrus-Erie broke up and sold its entire Construction Equipment Division. The remaining larger machines, classified as "mining and quarry shovels" along with the blast hole drills became the company's main products.

Many of the machines described in this chapter belonged to families of machines based on similar designs. After one successful machine was launched, others followed that were enlarged or modified into new models but retained the same basic design features of the pioneer machine. To avoid repetition of text, these groups or families of machines are referenced in Appendix I. Although not specifically mentioned, many of these smaller machines were offered with electric motor drive as an alternative to internal combustion engines or steam power.

Machines in this chapter are arranged in numerical or alphabetical sequence in ascending order by machine size so a particular machine can be easily found. Also refer to the tables in Appendix III, for production numbers, manufacturing dates and machine capacities not included in the photo captions. Because certain machines enjoyed a very long manufacturing run and updates often caused their weight to increase, the weights quoted are average or that quoted at the time of introduction.

A. The model A was inherited from the Erie Steam Shovel Company purchased by Bucyrus Company in 1927 to form Bucyrus-Erie Company. Formerly the Ball Engine Company, it had been making very successful and popular small revolving steam shovels since 1914. The model A was the smallest of Erie's five-model range, having a capacity of 1/2 cubic yard. It was available on traction wheels or crawler tracks and weighed 13 1/2 tons or 18 tons respectively.

AA. This model was a special shovel derived from the model A and designed for tunnel excavation. Powered by compressed air, it featured a short cambered boom at a fixed angle, short handle, and short tail swing, built tough for underground abuse. The AA was a popular tunnel shovel, and one of the last Erie-designed models to be sold as a Bucyrus-Erie. AAs were shipped up to 1935.

B. The Erie B was the first and most popular of the Erie steam shovels and, because of its longevity and robustness, many survive today in preserved condition. The 3/4-yard B could be mounted on traction wheels, crawler tracks or on rails as a railroad ditcher, and was available with a variety of boom and handle lengths as well as dragline equipment. Operating weight was approximately 22 tons.

B-2. In 1925 the Erie B was upgraded to the B-2 "Dreadnaught" with a slightly larger hoisting engine and capacity increased to 7/8 cubic yard. The high-lift boom version was used to strip overburden in mines or load trucks at ground level from a deep basement position. The long dipper handle attachment was used to dig trenches the old-fashioned way from a position over the trench.

B-3. The Erie B-2 steam excavator continued as a Bucyrus-Erie product after the two companies merged in 1927. In 1929 it was upgraded to the B-3 with a modern chain-driven undercarriage and weight increased to 32 tons. Bucyrus-Erie B-3s were sold up to 1936, marking the end of the steam shovel era. Total number of all B models shipped amounted to over 3,700 units.

GA-2. With an eye on its competition and realizing the days of steam power for excavators were numbered, Erie built a new type of machine called the "Gas + Air," said to perform with the finesse and control of a steam shovel but to run with the economy of a gasoline shovel. The 1-yard excavator was powered by a Waukesha 80-horsepower gasoline engine that drove an air compressor to power the air-operated motors for swing and crowd motions. The hoist and propel were directly driven from the engine through clutches.

GA-3. Bucyrus-Erie took over the Gas + Air concept and pursued it intently for a few years. The GA-2, originally appearing in 1925 as an Erie machine, was upgraded in 1930 to the Bucyrus-Erie GA-3 at 1 1/4-yard capacity and working weight of 36 tons, some 500 lbs. heavier than its predecessor. Power for the GA-3 came from a 95-horsepower Wisconsin gasoline engine. The complicated machinery in the Gas + Air machines (engine, compressor, and both air and mechanical systems) scored against them when it came time to rationalize the range in the early 1930s.

DA-2. In 1927 Erie built one diesel version of the GA-2 called the DA-2, with the same dimensions and rated capacity as the gasoline version. Then after the merger it continued as the Bucyrus-Erie DA-2 with a heavy 4-cylinder Atlas-Bucyrus-Erie diesel engine developing about 80 horsepower at a low 675 rpm. This added some 5,000 lbs. to the machine's weight. In 1930 the DA-2 became the **DA-3** but was discontinued the following year.

D-2. The D-2 and E-2 were Bucyrus-designed excavators of 1-yard and 1 1/4-yard capacity. They were introduced in 1927, the same year Bucyrus acquired Erie, so no doubt their model designation was intended to conform to the Erie nomenclature. Shovel attachments were of the twin-handle automatic rope crowd, so changeover to dragline involved no change in deck machinery. The D-2 shovel is illustrated here.

E-2. The D-2 and E-2 were only sold as crawler machines with gear driven undercarriages, and powered by slow-running Bucyrus-Atlas diesel engines. Initially they were 3-cylinder engines, the D-2s having a 7 1/2-in. bore and 10 1/2-in. stroke running at 400 rpm, while the E-2's engine ran even slower at 350 rpm with a 9-in. bore and 12-in. stroke. Due to model rationalization in the early 1930s, the E-2 was discontinued in 1930, while sales of the D-2 hung on until 1933.

TE-5. The TE-5 was a special electrically powered 1 1/4-yard tunnel shovel with an operating weight of 35 tons. Dipper handle length was 9 feet, and boom lengths of around 10 feet were offered to suit tunnel dimensions. Electrical components and propelling machinery were mounted on one end of the base, which could be mounted on rail wheels or crawler tracks, while the front end, cab and hoisting machinery were mounted on a revolving turntable at the other end. Several TE-5s were sold between 1931 and 1935.

1020. The "10-series" was a new line of excavators designed by the Erie engineering team following unsatisfactory sales of the Gas + Air machines. The first machine, the 3/4-yard 1030, was ready at about the time Erie was acquired by Bucyrus; consequently it was launched as a Bucyrus-Erie product. Shown here is the 1/2-yard 1020 backhoe, the smallest in the line. Introduced in 1929, it was equipped with a Waukesha gasoline engine rated at 54 horsepower.

1025. The 1025 rounded out the six-model "10-series" line of excavators in 1930. In shovel form it weighed 24 tons and was rated at 5/8 cubic yard. All the "10-series" machines had a very neat machinery deck layout as the picture shows. There were only two main shafts driven by the engine. The rear shaft carried the reversing clutches for swing and travel, and the front shaft carried the two main drums with their associated clutches and brakes.

1030. In 1928 the trailblazer for the "10-series" excavators was unveiled. This was the Bucyrus-Erie 1030 equipped with a Waukesha 94-horsepower gasoline engine and tipping the scales at 30 tons as a shovel. The 1030 turned out to be the biggest seller in the 10-series, with 227 leaving the factory in the five years prior to 1933. Initially 10-series excavators were gasoline powered, but soon diesel and electric power, like this 1030 electric shovel, were offered.

1030. By 1930, the "10-series" excavators had all been released. That was also the year Ruston-Bucyrus was established in England. The 1030 was chosen as one of the very first American-designed excavators to be built in the Lincoln plant. Another model, the 5/8-yard 1025 was also built at Lincoln. A Ruston-Bucyrus 1030 "drag shovel" is shown here.

1035. The 7/8-yard 1035 made an entry in 1929. Its Waukesha engine put out 84 horsepower running at 1,000 rpm, and as a shovel it weighed 33 tons. Principally the "10-series" shovels were sold with the twin-arm automatic rope crowd, but a chain-driven crowd, requiring an extra reversing shaft on the machinery deck, was also offered. One of these is shown here.

1060. The largest in Bucyrus-Erie's "10-series" hardly got its start before falling victim to model rationalization in the early 1930s. The 1 1/2-yard machine didn't reach beyond prototype stage, but one shown here is doing stalwart duty as a dragline. The 10-series machines were all discontinued by 1934 but many of their features could be found in Bucyrus-Erie's subsequent models.

2030. The 2030 was a 5/8-yard special machine based on the 1030 and built for railroad work. Its swing angle could be limited by applying a series of stops, manually set from the ground, which prevented swinging the crane over adjoining track, allowing work to proceed while trains passed. The rear end was tapered to prevent overhang, and the counterweight was moved further back to maintain balance. The Pennsylvania Railroad purchased 33 of these machines in 1929.

10-B. Launched in 1934, this smallest of the long line of Bucyrus-Erie excavators turned out to be one of the greatest success stories in excavator engineering. Affordable by the small contractor, the 3/8-yard 10-B was able to mechanize excavation work previously done by hand labor. With its simple design, no air lines, hydraulics or complicated mechanisms, reliability was the result. The early cab version is shown here with straight rear panels as opposed to the curved version that appeared about 1949.

10-B. A true universal machine able to operate as a shovel, backhoe, dragline, skimmer, clamshell, crane and pile driver, the 10-B weighed around 10 tons. It boasted a two-speed travel arrangement and could be steered "like a tractor" through friction clutches operated by a crank lever in the operator's cab. Gasoline or diesel engines rated at 38 to 52 horsepower by Hercules and Buda were installed. Shovel type was the single stick rope type.

10-RB. The need for such a small excavator originated in Great Britain soon after Ruston-Bucyrus Ltd. was established at Lincoln in 1930. Thus the jointly designed 10-B/10-RB machines were released simultaneously in 1934. The 10-RB was identical to its American counterpart except for the Ruston engine and a square-looking cab known as the "Lincoln" cab that gave more room inside for maintenance.

10-RB. The square cab with sloping upper edges was known as the "Continental" cab because it was designed to ship on a railway flatcar through the tunnels of the European Continent. The 10-RB was initially fitted with the Ruston 3-cylinder JRN diesel engine that gave 33 horsepower at 1,000 rpm. A Fordson gasoline-paraffin engine was also offered.

10-RB. It wasn't until 1955 that the Bucyrus-Erie cab with sloping left side appeared on the 10-RB. At that time it also received a lattice dragline boom to replace the former channel boom. Air control for steering and braking was added in 1961. The American 10-B was discontinued in 1959 after sales of 2,428 units, but the 10-RB continued to enjoy a buoyant market in Europe for another decade, notching up sales of 7,625 units. More 10-RBs were built in Australia by subsidiary company Ruston & Hornsby Pty. Ltd.

11-B. The 11-B was sold as a truck-mounted universal excavator, although a prototype was built as a crawler-mounted unit. It had the same general machinery arrangement as the 10-B but featured a double-row ball bearing swing circle bearing, nylon bushings in the operating linkages, and a low-profile house with isolated full-vision cab.

11-B. The 11-B truck crane was rated at 10 tons capacity, or 3/8 cubic yard as a shovel, hoe, dragline, or clamshell. The carrier was a 6x6 heavy-duty chassis equipped with a 120-horsepower Chrysler gasoline engine, while the upper could be powered by a Continental diesel or gasoline engine. The 11-B was made from 1958 to 1962 with a higher-horsepower Series II machine appearing in 1960. There was also an 11-RB version made at Lincoln with the same Ruston engine as in the 10-RB, and a 98-horsepower Perkins diesel in the carrier.

14-B Steam Shovel. The first fully revolving "construction size" excavators Bucyrus built were the 14-B and 18-B steam shovels. They were the first machines to be built in the brand new Evansville, Indiana, plant beginning in 1912. The 14-B shown here was rated at 5/8 cubic yard and weighed 21 tons in operation. It is immediately identified by its single crowd arm as opposed to the twin arm design on the 18-B.

14-RB. Ruston-Bucyrus released the 1/2-yard 14-RB and 5/8-yard 16-RB excavators in 1934. Their physical dimensions were the same; they differed only in weight and engine power. The 14-RB shovel weighed 17.5 tons and its Ruston diesel or Dorman gasoline engines were rated in the 44 horsepower class. A 14-RB dragline is shown.

14-B. Bucyrus-Erie produced the 14-B from 1960 to 1967. It was based on similar lines to, and was the company's replacement for, the 10-B. Rated at 1/2 cubic yard, the 14-B was physically the same size as the 11-B superstructure but with a slightly higher horsepower rating. It utilized air-controlled propel clutches and brakes. A "Super Crane" version allowed the 14-B to lift 15 tons at 10 feet radius. In the early 1960s, there was also a 14-RB of this design built by Ruston & Hornsby Pty. Ltd. in Australia.

15-B. The 1/2-yard 15-B was one of the most popular of Bucyrus-Erie's small excavators, and a favorite with the U.S. armed forces. Produced from 1940 to 1968, a 5/8-yard Series II was introduced in 1961. It followed the same machinery layout and front-end design as the 10-B. Gasoline or diesel engines in the 50-horsepower class were offered, initially Buda and Caterpillar, later Hercules and GMC options.

16-B. The 16-B was offered as a new 1/2-yard excavator in 1932. Fully convertible to all front ends, it was powered by a 54-horsepower Waukesha gasoline engine. Its machinery layout was the first of a new design, and the forerunner of a large number of Bucyrus-Erie excavators to follow, beginning with the 10-B. The 16-B was superseded by the 15-B in 1940.

16-RB. This is the Ruston-Bucyrus version of the 16-B and is almost identical to its sister 14-RB, except the engine was 10 horsepower greater, and weight about one ton heavier. The 14-RB/16-RB machines utilized an internal swing gear with hook rollers, and dual hoist drums on one shaft. Rated as a 5/8-yard excavator, the 16-RB shovel employs a chain-driven twin-handle crowd setup as shown.

17-B. The rare 17-B was another 1/2-yard machine made while the 1/2-yard 16-B was in production. The 15-ton 17-B incorporated a more refined machinery layout like that of the 10-B: drums lowered to deck elevation, and simplified control lever linkages neatly positioned below the deck. The 17-B was only made from 1936 to 1938 but it grew into the 5/8-yard 18-B with the same crawler undercarriage, slightly more horsepower, and weighing one ton heavier than the 17-B.

17-RB. Both the 17-RB and 19-RB excavators of 1/2-yard and 5/8-yard capacity were announced in 1937. The 17-RB carried either Ruston gasoline or Dorman diesel engines of 44 horsepower. The 17-RB skimmer in the picture is fitted with a "Lincoln" cab. Since the specifications and pricing of the 17-RB and 19-RB were so similar, customers nearly always bought the 19-RB, so the 17-RB was dropped by 1941.

18-B Steam Shovel. Rated at 7/8 cubic yard and with a weight of 25 tons, the 18-B was the larger of the two steam shovels which Bucyrus built to enter the small revolving shovel market in 1912. For the first few years both the 14-B and 18-B were mounted on traction wheels or rails, but it wasn't long after the advent of crawler tracks on excavators that tracks began to appear on both these models. The 18-B with its distinctive twin crowd arms is illustrated here.

18-B. The 5/8-yard 18-B replaced both the 1/2-yard 17-B and the 5/8-yard 19-B in 1938, leaving the 15-B to serve the 1/2-yard market. All these models looked very similar in appearance with physical dimensions varying by only few inches. Their gasoline and diesel engines were supplied by Buda, Hercules and Caterpillar, with power ratings ranging from 47 to 65 horsepower.

19-B. The 19-B was a 5/8-yard version of the 17-B, and both were made from 1936 to 1938. They possessed the same efficient machinery layout that first appeared on the 10-B, but of course beefed up in size. Power options were either a Wisconsin gasoline engine rated at 60 horsepower or a Caterpillar diesel at 58 horsepower. Operating weight as a shovel was 17 tons.

19-RB. Although the American 19-B was discontinued in 1938, the British 19-RB by contrast enjoyed a very long production run. Over 5,000 were sold up to 1966. Rated at 5/8-yard capacity it was initially fitted with Ruston gasoline or Dorman diesel engines of 54 horsepower, but the gasoline option was removed from later specifications. This 19-RB "drag shovel" sports a "Lincoln" cab.

19-RB. The original 19-RB "Lincoln" cab was later modified to the "Continental" cab with roof rounded at the sides for railroad shipment on flatcars. The last edition of the long running 19-RB was the standard "Bucyrus-Erie" cab which it gained in the mid-1950s. One of these is shown performing clamshell duties in a coal yard.

20-B Steam Shovel. This machine was one of the most successful steam machines from the Bucyrus stable. Introduced in 1922 and modeled after the 30-B steam shovel launched two years earlier, the 20-B replaced Bucyrus' first small revolving steam shovels, the 14-B and 18-B. Advertised as a new design "from boom point to ash pan" it was marketed as the "20-B Bullseye." Gasoline and diesel power options soon became available. The 20-ton, 3/4-yard excavator continued as a Bucyrus-Erie machine until 1928 when a total of 413 had been shipped.

20-B. Released at the same time as the 18-B in 1938, the 20-B was a 3/4-yard version of the same machine. Its larger capacity was supported by a slightly larger counterweight, larger standard undercarriage with 20 in. shoes, and more power from either a 71-horsepower Wisconsin gasoline engine, or a 67-horsepower Caterpillar diesel engine. The **20-RB** was also made at Lincoln, England, in 1940/41.

21-B. The 21-B embodied many features of the Erie-designed "10-Series" machines, still current when the 3/4-yard machine was introduced in 1931. Similarities included external gear swing rack with center pintle support (no hook rollers), spur gear engine power take-off, side-by-side hoist/drag drums on the same shaft, and shovel with twin handles driven by rack and pinion. The 21-B was the forerunner of many future Bucyrus-Erie excavator models of similar design.

21-RB. The British 21-RB was released simultaneously with its American counterpart—identical except for the usual engine change. (It was actually referred to as a 21-B.) The diesel machine utilized a 4-cylinder Ruston engine of 62 horsepower. The 21-RB shovel could be purchased with either an automatic rope crowd weighing 27 tons, or independent chain-driven crowd weighing 1,300 lbs. more. It was fully convertible to all the usual front ends including skimmer.

22-B. When the 22-B was announced in 1937 it was a special long-range dragline/crane version of the 20-B, having longer and wider crawlers and some extra counterweight. The resulting machine weighed 20 tons, some two tons heavier than the 20-B dragline. Customers liked the extra capability, so by 1941 the 22-B became a universal 3/4-yard excavator with full range of attachments. During World War II, all excavator models of 1-yard and under were discontinued except for the 10-B, 15-B and 22-B.

22-B. Bucyrus-Erie sold more 22-Bs than any other excavator. A major success in the excavating industry, it was used around the world in every conceivable application. The 22-B machinery layout shown here was the same as introduced in the 10-B in 1934. The 22-B Series II appeared in 1960 with 40 advertised improvements, and in 1964 the Series III arrived with further slight modifications and a more square-shaped cab. All told, 8,828 22-B's were shipped until the last one in 1976.

22-RB. The 22-RB joined the Ruston-Bucyrus roster of machines in 1950. Initially the 3/4-yard machine sported the "Continental" cab, like the one shown here loading from a quarry face in Scotland. From about 1955 the 22-RB, as well as the 10-RB and 19-RB, were fitted with the "Bucyrus-Erie" cab. The 22-RB carried the Ruston 4YEN engine of 66 horsepower. The 22-B was also manufactured under license in several overseas countries, including Canada, Brazil, Sweden and Japan.

22-RB. In an effort to compete with the myriad of hydraulic excavators entering the market, from 1963 Ruston-Bucyrus offered a hydraulic pivoting dipper for the 22-RB cambered-boom "dragshovel." This gave it the prizing "wrist" action and load retention of a hydraulic excavator. The 22-RB was also available in various forms of lifting crane, including "heavy-duty" variations with long and wide undercarriages, and lifting capabilities up to 11 tons. All were designed to broaden the market of the 22-RB.

22-RB icd (improved crane & dragline). In the progressive development of the well-established 22-RB excavator, the "icd" version was announced in 1965. This was a crane/dragline version with increased dragline rating to one cubic yard, and a crane rating of 16 tons at 10 feet radius. This was accomplished by redesigning the crawler mounting and counterweight arrangement, and utilizing high-strength steel in the boom chords. Optional air control for the main digging functions became available for the 22-RB in 1966.

22-RB. The success of the 22-RB continued long after its American 22-B counterpart had been discontinued. In its various forms of lifting cranes and excavators it continued to sell in healthy numbers. In the early 1970s, its engine options changed to the Dorman 6DA air-cooled diesel or the Detroit 4-71 water-cooled diesel, both of 75-horsepower rating. By the time the last new 22-RB excavator had left the Lincoln factory in 1994, a whopping 10,245 units had been shipped from there.

24-RB. When Ruston-Bucyrus took on the manufacture of the 21-B as a 3/4-yard machine in 1931, it became apparent that its robust "overdesign" could support a larger rating. So in 1938 it was replaced by the 24-RB of 7/8-yard capacity. A little more counterweight was added and a 77-horsepower Ruston engine replaced the former 62-horsepower model. The 24-RB tipped the scales at 30 tons, and remained in production until 1955. There was no American equivalent of this model.

25-B Steam Shovel. The 25-B was part of Bucyrus' first line of small revolving steam shovels. Built from 1914 to 1920, the 1 1/4-yard shovel was the third in the line after the 14-B and 18-B were introduced. Built in the new Evansville, Indiana, plant, the 25-B weighed 33 1/2 tons in operation, and was only available on a rail truck mounting as shown.

25-B. The modern 25-B was basically a 22-B with air control and upgraded to 1-yard capacity. House dimensions and engine options were the same, but the 25-B adopted the square-looking cab similar to later 22-Bs. Larger counterweight configurations could be mounted on the 25-B, and its standard undercarriage was 10 in. wider and 10 in. longer than the 22-B. The 25-B Series II with minor modifications and 1,600 lb. increase in weight appeared in 1964, and the Series III followed in production from 1968 to 1984.

25-SC. There was no 25-B Ruston-Bucyrus machine for general excavating, but the 25-SC (Super Crane) version was marketed from 1969 to 1977. It was a special lifting crane based on the air-controlled 25-B superstructure, but with extra counterweight and widespread undercarriage measuring 14 ft. 7 in. long, by 11 ft. 8 in. wide with 30-inch shoes. Its maximum lifting capacity was 67,000 lbs. at 10 feet radius.

27-D. The 27-D was a steam-powered crawler-mounted skimmer-type excavator of 1 1/2 cubic yards capacity. It found favor in surface coal mining where thin seams had to be separated from partings. A steam engine fixed midway on the boom thrust the bucket horizontally utilizing a rack-and-pinion drive. The machine weighed 33 tons in operation and 31 were sold from 1915 to 1922.

29-B. This was a 1-yard crawler excavator designed on similar lines to the 21-B. The 29-B was announced in 1937 and built for a 6-year period. Engine options were a Wisconsin gasoline of 98 horsepower, or a Caterpillar diesel of 92 horsepower. The chain crowd shovel weighed 33 tons and the machine was convertible for dragline, backhoe, dragline, clamshell or crane service.

30-B Steam Shovel. Bucyrus launched this pioneering steam shovel in 1920 after experience with its very first small revolving shovels, the 14-B and 18-B. Crawler, traction wheels or rail truck mountings were available, and a long boom for dragline or clam work was offered. The 33-ton, 1-yard 30-B continued as a Bucyrus-Erie machine until 1928 when a total of 328 had been shipped.

30-B. Soon after it was launched, the successful 30-B steam shovel was offered with diesel or gasoline power. The diesel engine was a 3-cylinder Atlas-Bucyrus having an 8 in. bore and 10 1/2 in. stroke, developing 55 horsepower at a low 360 rpm. The gasoline engine ran at 400 rpm with a 4-cylinder engine of 7 1/2 in. bore and 10 in. stroke. These engines required 2,000 lbs. of weight in their flywheels. The 30-B rope-operated crowd was of unusual design. (See 50-B for description.)

30-B. The "modern" 30-B was introduced in 1956 as a new 1-yard universal excavator. It followed the same design as the well proven 22-B but with larger dimensions and air control for all motions, one of the first from Bucyrus-Erie to be so equipped. The 30-B backhoe shown is a "hi-walker" equipped with a tractor-type undercarriage, a favorite for pipeline work often requiring machines to travel long distances over rough terrain.

30-B II. In 1964, the 30-B graduated to a Series II with increased ratings to 1 1/4 cubic yards, and an operating weight of 34 tons as a shovel, some 4,400 lbs. heavier than the former model. A variety of diesel engines was offered in the 110 to 124 horsepower range, and the "Supercrane" version could lift 50 tons at 10 feet radius. 30-B customers could choose from three crane boom types: heavy duty for heavy short-radius lifts, long range for booms up to 130 feet, and standard for excavating duties.

30-B III. The evolving 30-B turned into to a Series III in 1968 with the same capacity ratings as the former version. The 30-B Series III was replaced by the Series IV in 1968 with "Supercrane" ratings raised to 60 tons. A Series V machine appeared in 1981 boasting a ball bearing swing circle, and mainly sold as a lifting crane with capacities up to 70 tons in "Supercrane" guise.

30-RB. Ruston-Bucyrus introduced the 30-RB in 1956, the same time as its American partner. It was the identical machine except for the usual change of power unit, in this case the 5-cylinder Ruston 5YFN diesel of 97 horsepower. Early 30-RB/30-B machines revolved on six hook rollers running in a cast roller path that included an internal-tooth swing rack. The single-stick rope crowd shovel and tandem drums reflected their 10-B design family.

30-RB VI. Like the 30-B, the 30-RB graduated through a similar series of development stages. The Series II appeared in 1968. There was no Series III, but a Series IV arrived in 1971 with the more powerful Dorman 6LE diesel of 125 horsepower. A pre-loaded cross-roller swing bearing ushered in the Series V in 1982. When Bucyrus-Erie discontinued the 30-B in 1984, the 30-RB was far from finished. The picture shows the 30-RB Series VI with joystick controls that replaced the Series V in 1988 and continued until 1995.

31-B. Another stalwart steam shovel was the 1-yard 31-B introduced by Bucyrus in 1926. Shovel and dragline versions were offered as well as a long-stick sewer shovel, and a high-lift shovel for loading trucks on top of the cut. Operating weight as a shovel was 31 tons. The 31-B was only in production for two years but managed to sell a healthy 85 units.

32-B. The 1-yard 32-B was a larger brother to the 21-B. Both were made simultaneously and both shared the same design features such as external swing rack, twin drums and spur gear power take-off. A chain- or rope-driven crowd shovel was offered, and engine choices included a Wisconsin 105-horsepower gasoline, or an Atlas-Bucyrus 91-horsepower diesel engine. The 32-B shovel tipped the scales at 37 tons.

32-RB. Ruston-Bucyrus announced the 32-RB simultaneously with the 21-RB in 1931 (known as the 32-B and 21-B). They were almost identical to the American versions except for power units, and of similar design to the outgoing Ruston-Bucyrus 1030. Engine options were a 5-cylinder Ruston diesel of 88 horsepower running at 920 rpm or a 6-cylinder gasoline engine of 105 horsepower at 1,100 rpm. The dragline shown here has a single-motor electric drive, an option available for most universal machines of 2 cubic yards or under.

33-B. The 33-B replaced the 32-B in 1937 but retained its family design features of the former model. Its rated capacity was upped to 1 1/4 cubic yards, obtained from a slight increase in undercarriage dimensions, resulting in an overall weight increase of some 2,200 lbs. As a shovel, only the chain crowd was offered, and a Caterpillar 110-horsepower engine was standard on the diesel machine.

33-RB. The 33-RB was recognized by its distinctly sloping left front cab, in contrast to the very square cab of the former 32-RB which it replaced in 1938. With the 1 1/4-yard 33-RB's introduction, Ruston-Bucyrus vacated the 1-yard machine market until the 30-RB was announced in 1956. The 33-RB came with a 5-cylinder 110-horsepower Ruston diesel or a 117-horsepower gasoline engine. Production continued until 1956.

34-B. This was basically a 1 1/4-yard version of the 1-yard 32-B excavator, with similar design features and engine options. The 34-B's beefier design increased its weight to 44 tons and, in shovel form, only the more substantial chain crowd type was offered. The 34-B was added to the line in 1934 and continued up to 1937 when it, along with the 32-B, was replaced by the 33-B.

35-B. The 35-B was the largest of Bucyrus' first line of small revolving steam shovels. It was added to the line in 1914, and only offered as a 1 1/2-yard shovel. It was one of the first excavators to be mounted on crawler tracks, but rail and traction wheel versions were also available. The crawler version weighed approximately 50 tons, the rail version 42 tons.

37-B. This was one of the bevy of new excavator models Bucyrus-Erie released in 1931. The 1 1/2-yard 37-B, along with the 21-B, 32-B and 43-B released at the same time, were all derivations of the Erie-designed 10-Series machines, highlighted in the 21-B description. Initially the 37-B diesel machine was equipped with a Bucyrus-Atlas diesel engine, but by 1934 a 135-horsepower Caterpillar diesel became standard. In operation the 37-B weighed approximately 48 tons.

37-RB. Ruston-Bucyrus added the 37-RB to its line in 1932 and kept it in production until 1947, in readiness for the 38-RB. The 1 1/2-yard excavator was almost identical to the American 37-B except for engine power. The Ruston VQBN 115-horsepower model was the choice of power but a 121-horsepower gasoline engine was also offered.

38-B. One of the all-time long-running greats from Bucyrus-Erie, the 38-B came on the scene in 1945. It was the smallest of a family of larger construction-sized machines developed from the 54-B and extending up to the 88-B. Initially launched as a 1 1/2-yard excavator, it graduated into a Series II machine in 1963 with air control and 2-cubic-yards capacity. Standard power was a Caterpillar diesel of 126 horsepower. As a single-stick rope crowd shovel the 38-B weighed just under 49 tons.

38-RB. The 38-RB was recognized by its larger house, squared-off at the rear in contrast to the sloping rear end of the 38-B house. This change was to accommodate the larger Ruston 4RPHN V4 diesel engine rated at 132 horsepower and running at 900 rpm. A popular machine for Ruston-Bucyrus, it became the Series II with air control in 1965, and gained more engine options including Dorman, Cummins and GM. It was finally eclipsed by the all-British 51-60 in 1988.

41-B Steam Shovel. As far as steam shovels went, the 41-B was one of the most modern. Introduced by Bucyrus in 1926, it continued as a Bucyrus-Erie machine after the merger. The 1 1/4-yard excavator employed all the rugged features of a rock-loading quarry shovel: gear-driven undercarriage, heavy castings throughout, and oversized shafts and gears. The 41-B steam shovel weighed 46 tons, and when changed to a dragline it weighed 43 tons with a 50-foot boom.

41-B Electric Shovel. The electric version of the 41-B could be purchased in either a 2-motor or 3-motor configuration. The 2-motor style could operate as an automatic rope shovel or as a dragline with one motor driving the swing, the other driving the two drums geared together and controlled by mechanically operated brakes and air-controlled clutches. The 3-motor style was suitable for shovel operation with separate motors and controllers for hoist, crowd and swing.

41-B Tunnel Shovel. This is an example of one of the many shovels that Bucyrus/ Bucyrus-Erie built for special applications. Several of the standard models were adapted for work underground by utilizing special short booms and sticks, and short rear end radius. Steam shovels were particularly well suited for this work as most could operate on compressed air with minimal power conversion like this 41-B air-driven shovel.

42-B. The 1 1/2-yard 42-B was sold as either a steam shovel or electric shovel. It evolved from the 41-B which it replaced in 1930. Overall house dimensions were the same as the former model but the left side was cut away to improve operator visibility. A longer undercarriage and increased weight of 49 tons gave the 42-B its increased capacity. The electric version was available with Ward-Leonard control; the steam version was one of the last steam shovels Bucyrus-Erie offered on the market.

43-B. This was the largest of the bevy of excavators Bucyrus-Erie brought out in 1931, all the same family design as the 21-B. A solid universal excavator of 1 3/4-yard capacity, the 43-B was available with gasoline, diesel or electric power in the 135-horsepower class, and convertible to dragline as shown here. Initially, in common with all machines introduced in the early 1930s, the 43-RB had riveted boom and components. These components were welded in later machines when that technology became in vogue.

43-RB. The British version of the 43-B enjoyed a much longer production run than its American counterpart which was replaced by the 44-B in 1939. The shovel front end could be equipped either as a rope or chain crowd, the latter weighing up to 59 tons. Apart from electric power, the only engine offered in the 43-RB was the Ruston VPHN rated at 132 horsepower. Ruston-Bucyrus discontinued production of this model in 1955.

44-B. The 43-B was upgraded to the 2-yard 44-B in 1939 with little change in physical dimensions to the house and undercarriage. The Wisconsin gasoline and Buda diesel engines in the 43-B were replaced by higher horsepower models, both rated at 170 horsepower for the 44-B. Welded construction was used throughout, and overall weight reached 60 tons.

45-B. This machine was also based on the 43-B but was a special single-purpose dragline. It featured a widespread undercarriage measuring 19 ft. 5 in. long by 11 ft. 8 in. wide. With 33-inch treads, the bearing area was 92 square feet, some 77% more than the standard 43-B dragline, resulting in a low 9 psi on the ground. In addition, the center of rotation was offset by two feet so the longer end could be positioned to avoid "nosing in" while digging. The 45-B appeared in 1932.

46-B. The 46-B was another single-purpose crawler dragline emerging a year after the 45-B. With only minor changes from the former machine, its weight increased by 3,600 lbs. The 45-B and 46-B carried more powerful engines than their sister 43-B machine to cope with dragline buckets up to 2 1/2 cubic yards. Engine options available were the LeRoi RXS gasoline and a Buda diesel, both rated at 153 horsepower.

48-B. Bucyrus-Erie brought out the 2 1/4-yard 48-B gasoline/diesel crawler excavator in 1936. It was one of the last of the old Erie-design Group B machines, with features like external swing rack and center pintle mounting, but it boasted a modern welded boom and independent propel drive with separate reversing shaft and clutches. Thus the 62-ton 48-B was a hybrid stopgap machine until the first of a new family of machines, heralded by the 54-B, arrived in 1939. The 2 1/2-yard capacity 48-RB electric version with Ward-Leonard control was produced at Lincoln, England.

50-B Steam. Originally presented by the Bucyrus Company in 1922, the 50-B was one of the most famous and successful heavy-duty excavators produced by the company. It continued as a Bucyrus-Erie machine from 1927 to 1934. The versatile 50-B was convertible to dragline duty and could be ordered with steam, gasoline, diesel or electric power, and mounted on rails or crawler tracks. The steam machine weighed 58 tons and was conservatively rated at 1 3/4 cubic yards.

50-B Diesel. The 50-B diesel version had a Bucyrus-Atlas diesel positioned longitudinally in the rear of the house. The 4-cylinder engine had a 9 in. bore and 12 in. stroke, and ran at a slow 300 rpm. The diesel shovel employed a rope crowd unique to the 50-B and 30-B machines. The shipper shaft was driven in either direction by two ropes wound on it. One rope ran via the boom point to an auxiliary drum on the hoist shaft, the other rope ran through the dipper sheave to the boom point and then back to the main hoist drum.

50-B Electric. Like other 1920s era shovels, the 50-B initially was sold with a wooden house as standard, but this was substituted with an all-steel cab in later versions. The machine was also uprated to 2 cubic yards when it became a Bucyrus-Erie machine. The electric 50-B mostly employed the universally accepted Ward-Leonard DC control, although AC drives were optional. Later models of the 50-B tipped the scales at just over 70 tons. A total of 534 50-Bs were sold.

51-B. The 51-B was basically an upgraded version of the 38-B with capacity increased to 2 cubic yards. It arrived on the scene a year after the 38-B. Machinery layout was the same except its boom hoist was driven through a separate set of bevel gears. Standard power was a Caterpillar D17000 diesel rated at 170 horsepower at 1,000 rpm. As a single-stick rope crowd shovel it weighed just under 61 tons.

51-B II. By 1953 the original Caterpillar engine in the 51-B had been replaced by a Buda diesel of 171 horsepower. Then in 1960 Bucyrus-Erie raised the 51-B to a Series II machine with weight upped to 63 tons and, as shown here, a 2 1/2-yard dipper for the shovel version. The same engine was installed as before but now under the Allis-Chalmers name. The 51-B was discontinued in 1964.

52-B. The 52-B was a larger version of the 43-B/44-B machines but with 2 1/4-cubic-yards capacity. Appearing in 1930, its design originated in the 10-Series Erie machines, but it gained separate sets of reversing bevel gears and clutches for both swing and propel motions on the diesel version, and its main clutches were operated by compressed air. The DC electric Ward-Leonard shovel is shown. The 52-B was a heavy machine for its capacity, tipping the scales at 83 tons in shovel form.

52-RB. Both the American-built 52-B and the British 52-RB (actually known as the 52-B) could be purchased with either diesel or Ward-Leonard DC electric power. But Ruston-Bucyrus also offered a steam version. Still rated at 2 1/4 yards, the 52-B steam shovel nudged up to 90 tons in weight. The one shown was shipped in 1935 to Alpha Cement Ltd. in central England, and was the very last steam excavator to leave the Ruston-Bucyrus plant. It is now preserved in a museum in Leicester.

54-B. Replacing the 48-B in 1939, Bucyrus-Erie introduced the similarly designed 54-B but with internal swing rack and hook rollers. Rated at 2 1/2 cubic yards, it was the forerunner of a new family of machines in sizes up to the 88-B. Its distinctive widely spaced dipper sticks with rack-and-pinion drive gave it robust characteristics allowing contractors to assign the 54-B the toughest work. A Buda 6-DH diesel of 236 horsepower was initially installed.

54-B. The popular 54-B could be readily converted to dragline with booms up to 100 feet and, after 1958, to a backhoe. In 1960, the 54-B graduated to a Series II rated at 3-cubic-yards capacity. By that time its engine options were GMC or Cummins in the 210-horsepower class. Over the years the 54-B was continually improved, with weight progressing from an initial 70 tons to almost 85 tons by 1962. Well over 1,500 54-B and 54-RB machines were produced.

54-RB. The 54-RB was recognized by its larger house, squared-off at the rear in contrast to the sloping rear end of the 54-B house. This was to accommodate the larger Ruston 6RPHN V6 diesel engine rated at 210 horsepower and running at 975 rpm. The diesel and Ward-Leonard electric 54-RBs were produced in significant numbers up to 1973. Two from George Wimpey's extensive fleet are shown here on highway construction in 1965.

55-B. The 55-B was the largest of those machines developed from the 10-series Erie-designed machines featuring external swing rack, center pintle mounting, etc. Its machinery layout was the same as the 52-B but the excavator was rated at 2 1/2 cubic yards, and in shovel form it utilized a rack-and-pinion chain-driven crowd as opposed to the automatic rope crowd on the smaller model. The 55-B was launched in 1935 and then in 1939 both the 52-B and 55-B ceased production.

55-RB. The Lincoln-built 55-RB was furnished with a 6-cylinder Ruston engine developing 175 horsepower at 525 rpm. It required a larger house than the 55-B: 1 ft. 8 in. higher and 1 ft. 3 in. longer to accommodate the big Ruston engine. Working weight increased to 90 tons, some six tons heavier than the 55-B. Power options were diesel, AC electric or Ward-Leonard DC electric. Many were built as draglines, as the picture shows.

61-B. Introduced at the Chicago Road Show in 1963, the 61-B was the last universal crawler excavator designed by Bucyrus-Erie. It displayed the very latest in design features such as: double-row ball-bearing swing circle, full air control, and dual levers controlling torque converter and engine governor. With a GMC diesel of 201 horsepower, the 61-B weighed approximately 80 tons. A Series II machine appeared in 1967 with shovel capacity upped to 3 1/4 yards, backhoe to 3 1/2 yards, and weight to 84 tons.

61-RB. Ruston-Bucyrus introduced the 61-RB in 1967. This time it was the identical machine to the American 61-B; even the 208-horsepower Cummins NT-855 engine was available on both machines. The 61-RB's widespread undercarriage with removable crawler side frames made an excellent lifting crane. Various "Heavy Duty" and "Supercrane" versions were available with lifting capacities up to 112 tons at 15 feet radius.

62-B. The 62-B was a special lifting crane adapted from a 61-B for magnet dropball work in steel mills. A typical application with a two-part hoist line would include a single-sheave 15,000 pound hook block, 15,000 pound magnet, and 30,000 pound dropball. The larger Caterpillar D343T engine of 247 horsepower was fitted. Other enhancements included wide-spread undercarriage with heavy-duty crawlers, elevated cab, and a blower for the main clutch and brake. The 62-B was available from 1975 to 1984.

65-D. The 65-D was a special machine for dragline, clamshell and heavy-duty crane service. Introduced in 1976, it was the last construction-size crawler excavator released by Bucyrus-Erie. Features included a hydrostatic swing drive, hydrostatic motors to propel each crawler independently, and a ball-bearing swing circle. A 280-horsepower Cummins NT-855C engine provided the power. The 65-D was uprated to Series II (1979) and Series III (1981) with draglines up to 4 yards, and maximum crane duties to 100 tons.

71-B. Bucyrus-Erie didn't offer a 3-yard excavator until the 71-B arrived in 1954. It filled the gap between the very successful 54-B and the 88-B machines, so its basic machinery layout was similar to them. The 71-B shovel utilized full air control, and a rope-operated dual-crowd type as equipped on the 88-B. But it embodied the internal swing rack-and-hook roller design of the 54-B. The power plant installed was the GMC 6-110 diesel at 230 horsepower.

71-B. The dragline version of the 71-B carried a standard bucket of 3 1/2 cubic yards on a 70-foot boom. In 1960, the 71-B matured into the Series II, with an increase in shovel capacity to 3 1/2 cubic yards and a weight gain to 100 tons, some six tons heavier than the previous series. It gained another five tons in weight in 1967 when the Series III appeared, this time with a standard shovel capacity of 4 1/4 cubic yards.

71-B. By the time the 71-B Series III was produced, several diesel power plants were offered in the 250-horsepower class from Cummins, GMC and Caterpillar. The 71-B was also available as a 3-motor Ward-Leonard excavator, and could be equipped as a backhoe, able to dig down to 36 feet. The last 71-B was built in 1983.

71-RB. The 71-RB was produced at Lincoln, England, from 1961 until 1992. It was actually the 71-B Series II that was released but with a Ruston 240-horsepower engine. By the time the Series III arrived in 1971, a Cummins NT-855 was the standard power, and the machine's weight had crept up to just over 101 tons.

72-B. The 72-B was an electrically powered machine specially engineered for crane, magnet or dragline work in steel mills. It was equipped with the Ward-Leonard control system with separate DC motors for hoist, drag and swing. To facilitate a quick change from heavy lifting (slow line speed) to high-speed dropball work, the electric control was arranged to change the ratio of line pull to line speed at the flick of a switch. During the 1960s Heckett Engineering Co. purchased more than two dozen of these units.

73-B. Another special dragline/crane for slag recovery in steel mills based on the 71-B appeared in 1964. The 4-yard rated 73-B featured a single motor electric drive to operate the drums through the usual clutches and brakes. It featured a widespread undercarriage and a high strength alloy boom. Because of the highly corrosive conditions present in slag recovery, both the 72-B and 73-B electrical systems utilized corrosive-resistant hardware with special insulation and a heater for each motor. Heckett Engineering Co. purchased eleven 73-Bs.

75-B. This was the smallest of the "Quarry and Mine" shovels, heavy-duty Ward-Leonard electric machines built for rock work, combining the ruggedness of the old part-swing railroad shovel with the flexibility of the fully revolving types. Introduced in 1928, the 104-ton 75-B shovel carried a standard 2 1/2-yard dipper. Many of these machines found themselves loading coal in strip mines with dippers of greater capacity than standard.

80-B Steam. The Bucyrus-designed 80-B came on the scene in 1921 available with steam or electric power, and mounted on rail trucks or crawler tracks. Its design followed closely that of the successful 225-B stripping shovel, but while the 80-B could work as a stripping shovel with a 48-foot boom and 2-yard dipper, most were equipped with a regular 35-foot boom and 2 1/2-yard dipper for loading hoppers or rail cars. The 80-B tipped the scales at 108 tons depending on its configuration.

80-B Electric. As a crawler machine, the 2 1/2-yard 80-B was the largest yet to be so mounted. It embodied a rare four-crawler design with two crawlers attached to an equalizing beam to accommodate uneven ground. As a stripping shovel, the 80-B was the first to be equipped with crawler tracks. When equipped as a dragline the machine was known as the Class 80. (See Chapter 5.) An electrically powered 80-B is illustrated.

84-B. A small number of 84-Bs were shipped beginning in 1964, both as shovels and draglines. The 4 1/2-yard machine was a smaller version of the 88-B, cut back in weight and horsepower. Except for a slightly narrower undercarriage, its physical dimensions including boom and handle were the same as the 88-B. And its overall weight was 4 tons less than the larger machine.

85-B. Introduced in 1935, the 85-B was the last of the original five-model line of "Quarry and Mine" electric shovels. It replaced the 75-B which proved to be too small as a heavy-duty electric machine—only 28 were sold. Unusually sized at 3 1/4 cubic yards, the 85-B utilized heat-treated alloy steels and welded construction, keeping its weight to an efficient 117 tons. Interestingly, Bucyrus-Erie retained the 85-B in its line until 1957, some six years after the other quarry and mine shovels had been replaced by updated models.

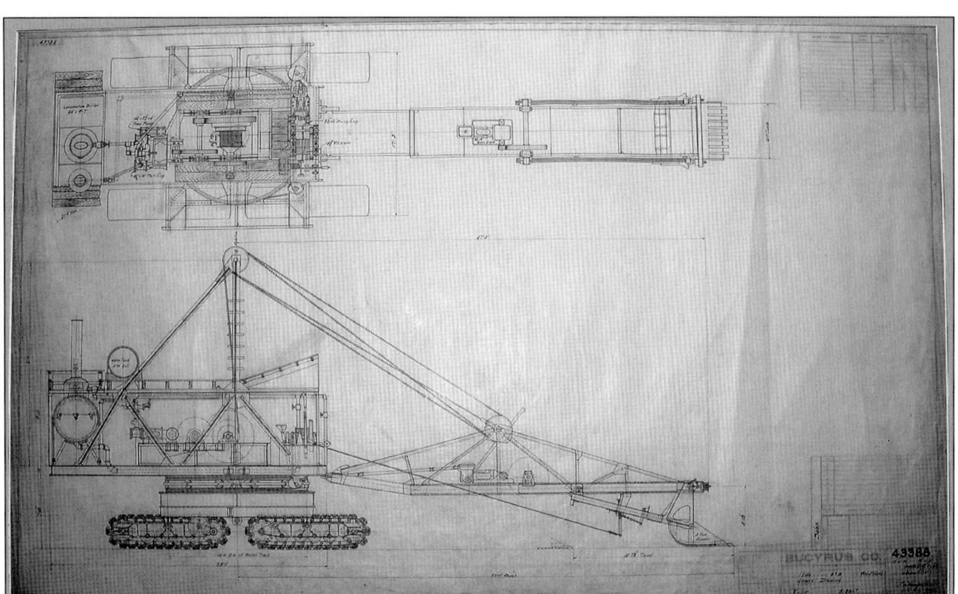

87-D. At least one of these special steam excavators was produced: serial number 2140 built in 1917. This drawing is the only information currently available. The 87-D was a much larger version of the 27-D skimmer and designed for excavating and loading thin coal seams. Its undercarriage was a four-crawler assembly, one of the earliest applications of crawler tracks on an excavator. The front-dumping skimmer bucket held 3 cubic yards and the steam-thrusting engine was mounted on the boom.

88-B. One of the most famous and successful excavators designed by Bucyrus-Erie was the 88-B. Launched in 1946 as a 4-yard excavator, it graduated to a 5-yard machine in its Series II form in 1960. With full air control, the 88-B was designed with a hook roller swing bearing, but utilized an external swing rack, differing from the other machines in its design family. The original engine installed in the 88-B was the 6-cylinder Cummins LRI 600 developing 264 horsepower.

88-B. For the 88-B series II, the Cummins 12-cylinder NVH-12 engine rated at 280 horsepower in conjunction with a torque converter was the standard motive power. The 88-B Series III followed in 1962, and then in 1968 the Series IV appeared, gaining another half-yard capacity. The 5 1/2-yard Series IV received a new raised cab and the Cummins KT-1150C engine at 365 horsepower in 1975. As a dragline, a live mast with pendant suspension supported the boom, as shown on these two 88-B draglines uncovering coal in Kentucky.

88-B Loader. In the twilight of its reign, and in a last attempt to prolong its life, Bucyrus-Erie revamped the 88-B Series IV as a 10-cubic-yard shovel. Announced in 1981, the "stretched" machine kept the same boom and handle as the standard machine, but gained more counterweight and more horsepower from the Cummins KT-1150C rated at 475 horsepower. Working weight also increased to just under 150 tons.

100-B Steam. A year after Bucyrus Co. introduced the 120-B, the world's first "Quarry and Mine" shovel, it began producing the smaller 3-yard capacity 100-B in 1926. Based on the same design but with reduced proportions, the 100-B weighed in at 128 tons. The early versions could be recognized by the forward-positioned A-frame which was moved further back by the time the machine carried the Bucyrus-Erie name. Ten of the 100-Bs were sold as steam shovels like the one shown here, all shipped prior to 1930.

100-B. During the 1930s, the ruggedness and somewhat overdesign of the 100-B, plus the introduction of welded components, allowed an increase to 4 cubic yards. The Ward-Leonard electric machine was equipped with a single vertical swing drive as opposed to two on the larger 120-B. Standard boom length was 32 feet. The 100-B was a very successful machine, staying in production until 1950 by which time its operating weight increased to 136 tons.

100-B Dragline. When a 100-B was required for dragline service it was usually ordered that way from the factory as it was a major job to convert it in the field. For dragline service a drag drum was added to the deck machinery together with associated support structure, gearing, clutches and brakes. Also a power-driven boom hoist was added, not necessary for the shovel boom as that was suspended at a fixed angle. The 100-B dragline carried buckets up to 4 1/2 yards on a 60-foot boom or 2 yards on a 90-foot boom.

100-RB. While the American-built 100-B moved up from a 3-yard to a 4-yard machine over time, the British-built 100-RB, introduced in 1933 was designated a 3 1/2-yard shovel from the outset. And it remained at that capacity until its demise in 1954. With upgrades, its operating weight increased from 138 tons to 162 tons during its production run of 83 machines made at the Lincoln factory.

110-B. The 4 1/2-yard Ward-Leonard electric 110-B shovel/dragline replaced the 100-B in 1950. It was a totally new design, one that Bucyrus-Erie would adopt for its quarry and mine shovels for the next four decades. The design embodied features proven on the larger stripping shovels such as a rope crowd shovel with tubular dipper handle, crowd machinery mounted on the revolving frame, two-part boom supported midway from the gantry, and advanced electrics. The 110-B was perfectly sized to match 20- to 50-ton haulers popular in the 1950s and 1960s.

110-B Dragline. Like the 100-B, most 110-B draglines were ordered so equipped from the factory. If converted in the field, the deck-mounted crowd machinery was not utilized, but an additional hoist drum with associated electric motor drive was installed behind the hoist drum on the deck. The shovel hoist drum then became the drag drum. Buckets up to 5 1/2 cubic yards could be carried on an 80-foot boom, although 4 1/2 yards on a 100-foot boom was standard.

110-RB. The 110-RB built in the U.K. replaced the 100-RB after 1954 and continued in production until 1981. Ruston-Bucyrus shipped 151 units (compared to 92 by Bucyrus-Erie) with approximately two-thirds exported to 25 countries. The 110-RB Ward-Leonard shovel carried a 4 1/2-yard dipper on a 36-foot standard boom and tipped the scales at 170 tons. The dragline version added roughly another 10 tons.

110-RB Dragline. The dragline version of the 110-RB was a popular machine in the U.K., especially when equipped with a self-contained diesel-electric power supply. Freed from its trailing cable, the diesel-electric shovel or dragline found mobility in many of the U.K.'s smaller coal mining operation where electric supply was not warranted. The 110-RB dragline in standard form swung a 4 1/2-yard bucket on a 100-foot boom. A Ruston-Paxman 12-RPHN.II 12-cylinder diesel engine rated at 528 horsepower was installed in these machines.

115-B. Using a magnet with a drop ball to break up slag in steel mills is one of the most grueling tasks for any crane. The load picked up is unknown, and the instant release of each load imposes severe strains on the machine. Enter the 115-B, a beefed-up 110-B reinforced in all stressed areas, and with long and wide undercarriage. Another steel mill special was the **114-B**, a lighter version with standard undercarriage. The 115-B was 28 tons heavier, and the 114-B 6 tons heavier, than the standard 110-B with the same-length 80-foot boom. The 115-B shown worked in Wales for Heckett Engineering Co.

120-B. The innovative 120-B was the world's first "Quarry and Mine" shovel, a heavy-duty Ward-Leonard electric machine built with the ruggedness of the old part-swing railroad shovel, yet with the flexibility of a fully revolving shovel. Initially the 120-B was rated at 4-cubic-yards capacity and tipped the scales at 168 tons. It was ruggedly built with massive castings and riveted construction. It featured wide-spaced dipper handles with rack-and-pinion drive and motor mounted on the boom. Like the 100-B, the early shovel version of the 120-B could be recognized by its forward-positioned A-frame.

120-B. When electric welding came of age in the mid 1930s, both the 100-B and 120-B shovels graduated from riveted construction to electric-welded construction. While retaining their rugged steel castings for carbody and revolving frames, their welded booms and other assemblies permitted increased dipper capacities to 4 yards and 5 yards respectively with no significant gain in overall machine weight. The 120-B revolutionized mass excavation methods and is the machine recognized as the forerunner of today's high-capacity Bucyrus electric mining shovels.

120-B Dragline. Some 120-Bs were sold as draglines which carried a maximum bucket of 5 cubic yards on a 70-foot boom. While conversion from shovel to dragline was possible, it was seldom done in the field because of the amount of work involved. Additions on the deck included a drag drum, boom hoist machinery, associated gearing, and modified operator's controls to work the clutches and brakes. The successful 120-B sold 303 units, four of which were steam machines. Two of these were sold to the New York Trap Rock Company in 1939, likely the last steam shovels built in America.

120-RB. The British-built 120-RB wasn't introduced until 1948, some 23 years after its American cousin. At 5-cubic-yards capacity both machines were almost identical, but slight cosmetic and mechanical changes added some 10 tons to the 120-RB's weight. The 120-RB's short production run ended in 1954 by which time Ruston-Bucyrus had placed a respectable 30 units in service.

150-B. Hot on the heels of the 110-B's launch in 1950, Bucyrus-Erie followed up with the heavier 150-B excavator a year later. Embracing all the design advantages of its sister machine—two-part pinned boom, single tubular handle to relieve torsional stresses, twin hoist ropes, crowd machinery deck-mounted close to center of rotation to minimize inertia—the 150-B Ward-Leonard turned out to be another well-received electric excavator in the 6-yard class. With a weight of 222 tons, it was a perfect match for popular 35- to 65-ton haul trucks.

150-B Dragline. Like the 110-B, the 150-B was offered as a 6- to 7-yard dragline and presented similar challenges to change from shovel to dragline in the field. Operating weight was listed as 243 tons with a 120-foot boom. The 150-B dragline was equipped with identical hoist and drag motors, each of 375 horsepower, and two swing motors of 88 horsepower. The 150-B remained in production until 1975 when it was replaced by the 155-B.

150-RB. Ruston-Bucyrus built the 150-RB from 1958 until 1983. The 150-RB carried shovel dippers of 6 or 7 cubic yards on a standard 37-foot 6-inch long boom, and weighed 237 tons. As a dragline, booms from 90 to 130 feet were supplied. A long-range shovel was also offered with 45-foot boom and 5-yard dipper. Ruston-Bucyrus shipped 125 150-RBs (compared to 151 150-Bs by Bucyrus-Erie), and while the majority worked in the U.K., others were exported to 22 countries.

155-B. The 155-B Ward-Leonard electric shovel was physically similar in size and electrical horsepower to the 150-B which it replaced. But it received engineering upgrades increasing its weight to 284 tons, allowing it to carry a standard 9-yard dipper. These specifications were up from 242 tons and 7-yard dipper on the last-series 150-B machines. Most of the 155-Bs were sold as coal-loading shovels equipped with 52-foot booms and dippers from 13 to 15 yards capacity. A single **155-BI** was sold in 1982 with its weight boosted to 294 tons and carrying a standard 12-yard dipper.

170-B. The 170-B was the largest in the first family of electric quarry and mining shovels Bucyrus Company initiated in 1925 with the 120-B. Introduced in 1929 as a Bucyrus-Erie machine, the 170-B was similar to its smaller brothers but with beefed-up proportions. The 170-B carried a 6 1/2-yard dipper on a 34-foot boom and weighed in at 244 tons. The massively proportioned shovel found its way into the toughest rock excavation projects of the day, including 33 to the iron ore mines of Minnesota and 19 to the anthracite mines of Pennsylvania.

182M. This is the first cable excavator (and the smallest) so far described in this chapter that is still currently available from Bucyrus International, Inc. The 13-yard (standard), 364-ton Ward-Leonard electric shovel remains from the former Marion line that Bucyrus acquired in 1997. The robust and reliable 182-M, with twin crowd arms driven by rack-and-pinion, continues as a Bucyrus machine. A significant number have been sold in India with local manufactured content for surface coal mining.

190-B. This was the last and largest of three new electric quarry and mining shovels Bucyrus-Erie released in the early 1950s to replace the former models designed in the 1920s. The 190-B replaced the 170-B in 1952. The same design as the successful 110-B and 150-B models but with expanded proportions and increased electrical power, the 190-B carried a standard 9-yard dipper and tipped the scales at just under 300 tons in working order.

190-B Dragline. Some 190-Bs were operated as draglines and, like their smaller brothers (110-B and 150-B), an additional drum and associated motor drive were necessary when converting from a shovel. Dragline booms from 100 to 130 feet were available. The machine illustrated began life on the Tarbela Dam job in Pakistan, then went to Saskatchewan, Canada, as a dragline with 110-foot boom, and was later converted to a 12-yard coal-loading shovel.

190-B. Over its long life from 1952 to 1976, the 190-B benefited from several upgrades but its sound basic design and electrical horsepower remained the same. The Ward-Leonard electric system controlled one 500-horsepower motor, one 125-horsepower crowd motor and two 100-horsepower swing motors. The machine illustrated sports a raised cab and is one of the last four built. These were delivered to the Highvale coal mine in Alberta, Canada, equipped with 15-yard coal dippers and are still working at the time of writing.

195-B. When the 195-B Ward-Leonard electric mining shovel was introduced in 1968 it was not a replacement for the 190-B which stayed in production until 1975. Both machines bore the same physical dimension, but the 195-B gained 26% in dipper pull and some 26 tons in overall weight, allowing its standard dipper to increase to 12 cubic yards. In 1978 Bucyrus-Erie commenced building the **195-BI** with electric motor propel drive in the lower works, increased weight to 394 tons, and standard dipper boosted to 16 cubic yards.

195-B (British). Ruston-Bucyrus commenced building the 195-B in 1974. It was never officially designated an "RB" machine, probably because its specification was identical to its American counterpart in every detail except for a few external cosmetic modifications. Forty-four were shipped from the Lincoln factory up to 1985 with thirty sold to customers for work in the U.K.

200-B Dragline. This machine was essentially a dragline version of the 195-B. It carried some 18 tons more counterweight than the 195-B, boosting its weight to 375 tons, and it swung a 12-yard bucket on a 100-foot boom. Electric power consisted of identical hoist and swing drives of 640 horsepower each, and two 130-horsepower swing motors. Four of these machines were sold to Morocco for phosphate mining.

210-B. The 210-B was a diesel-electric dragline of unique design. The hoist and drag drums were driven through eddy-current clutches, the input shafts of which were chain driven at constant speed by a rare White Superior model 40-SX-8 diesel engine putting out 900 horsepower at 900 rpm. The operator's hoist and drag levers controlled the excitation within the clutches and hence the speed and direction of each drum. The diesel engine also drove a generator to power a single swing motor. The 275-ton dragline worked with a 9-yard bucket on a 100-foot boom.

270-B. Produced over a short 2-year period beginning in 1960, the 270-B was a milestone in shovel design. It was Bucyrus-Erie's first step toward designing large electric mining shovels resulting in the giant machines of today. The 270-B was the industry's first static-controlled excavator utilizing silicon-controlled rectifiers to vary intensity and direction of the main generator excitation current. The shovel boom was a single-piece structure, departing from Bucyrus-Erie's two-part design. The standard 270-B weighed approximately 380 tons and its dipper held 12 cubic yards on a boom 45 feet 6 inches long.

280-B. In 1962 Bucyrus-Erie replaced the 270-B electric mining shovel with the higher capacity 280-B. With enlarged dimensions, including longer and wider undercarriage and larger diameter swing circle, it retained all the advanced features of the former model, including the single-piece boom and tubular dipper handle. This 280-B, owned by Kaiser Steel Corporation, is loading a 100-ton Dart articulated hauler at the Eagle Mountain Mine, California.

280-B. The highly successful 280-B could easily handle a 15-cubic-yard dipper on a 50-foot boom. Over its manufacturing life span of 20 years, the 280-B was continually upgraded. The later machines weighed 495 tons, some 23% heavier than the first 280-Bs to leave the factory. A typical upgrade example is the substantial boom on this late-version 280-B working for Kennecott Copper Corporation in Arizona. It also has a left-hand cab at the customer's request.

290-B. Although the 290-B was announced in 1976, the first was actually shipped in 1979 as a **290-BI** model with a 20-yard dipper and 47-foot boom. Others were sold with a 13-yard heavy-duty dipper for iron ore mining. The 290-B1 featured a propel motor in the lower works and embodied all the design features of the already proven 295-B. Working weight was 510 tons. In 1983 the 290-BI graduated to the **290-BII** with standard boom increased to 50 feet and an AC controlled frequency drive system called "Acutrol."

295-B. In 1972 Bucyrus-Erie introduced the 295-B, its largest mining shovel to date. It was the first of a long line of big modern electric mining shovels that continue today in their updated form. Average operating weight was around 600 tons depending on bucket/boom configurations. Dippers ranged from 13 yards for iron ore to 35 yards for coal, the latter shown here—one of seven 295-Bs employed at the Belly Ayr coal mine, Wyoming.

295-B. The 295-B Ward-Leonard electric shovel boasted a 375-horsepower electric motor in the lower works for propelling the machine through air-operated disc clutches. Other motors included a main hoist motor of 800 horsepower, a 195-horsepower crowd motor, and two 195-horsepower swing motors. The rugged single-piece boom followed the design proven on the successful 280-B.

295-BI. Beginning in 1980, Bucyrus-Erie boosted the capacity of the 295-B to 27 cubic yards and with other upgrades called it the **295-BI**. Then in 1983 the **295-BII** appeared with standard boom extended to 53 feet. This machine featured squirrel-cage AC motor drive with a controlled frequency drive system patented by Bucyrus-Erie as "Acutrol" and first applied on the 395-B shovel introduced a year earlier. *Photo courtesy of Gary Middlebrook*

295-BII Diesel. Interesting adaptations of the 295-B shovel are two diesel-electric machines operated by TriStar Mining in Maryland, believed to be the world's largest diesel-powered cable excavators. The machine illustrated carries a Cummins QSK60 diesel engine developing 3,000 horsepower, effectively replacing the normal AC synchronous motor to generate the DC current required to power the hoist, crowd, swing and propel motions. Dipper capacity is 35 cubic yards.

395-B. In 1980, Bucyrus-Erie continued its long tradition of technology leadership by shipping the first electric mining shovel equipped with AC electric motors and controlled-frequency electric drive "Acutrol" system patented by Bucyrus-Erie. The 34-yard 395-B offered increased reliability and lower maintenance by eliminating brushes and commutators in the motors. The crawlers featured bogie-mounted lower rollers that oscillated over pit floor obstructions. The first 395-B was shipped to the Anamax Mine near Tucson, Arizona.

395-B. Syncrude Canada ordered three 966-ton 395-B shovels in the mid-1980s for their oil sands operation in Alberta. The main AC electric drive comprised a 1,500-horsepower hoist motor, a 390-horsepower crowd motor and two 330-horsepower swing motors. Lower works propel was provided by a single 1,070-horsepower motor with air-operated clutches for steering. In 1989 the 395-B was superseded by the 395-BI and 395-BII shovels. These upgraded machines boasted standard dipper capacities of 39 and 43 cubic yards with weights increased to 1,034 and 1,062 tons respectively.

395-BIII. In 1995 Bucyrus-Erie commenced manufacture of the 395-BIII, identical to the 395-BII except it boasted a two-motor lower works propel drive, eliminating the clutches. Like the Series II it also embodied the latest in shovel technology including a completely computerized digital control system with in-built diagnostic capabilities. It also carried a standard boom of 64 feet in length and nominal dipper of 44-cubic-yards capacity. Weight was upped to 1,144 tons on this model.

495-B. The first of the 495-series electric shovels was introduced in 1990. Designed on similar lines to the 395-B it embodied all its advanced mechanical features and AC motors with "Acutrol" variable frequency control. Standard dipper of 53 cubic yards was specified with a 64-foot boom, and weight increased to 1,228 tons. Bucyrus enhanced the 495-B series through the **495-BI** (1996) and **495-BII** (2000) with added features including dual motor independent planetary propel and a planetary hoist drum drive with twin output pinions.

495HR. The 495 H-Series, first appearing in 2002, represents the largest and newest electric mining shovels available from Bucyrus International, Inc. at the time of writing. All models share similar design, with Siemens AC motor electric drive system and 67-foot booms. The **495HD** (Heavy Duty) was derived from the 495-BII with 90-ton dipper payload and improved AC drive system. The 110-ton payload 495HR (Hard Rock) version utilizes an undercarriage derived from the former Marion 351M/595B design equipped with narrow (79-inch) crawler shoes for rock applications.

495HR. The current AC drives in the H-series Bucyrus machines incorporate the latest in electric drive technology. They evolved through three different stages: the original as found in the 495-B, an improved GTO (gate turn-off) version, and the latest extremely reliable IGBT (insulated gate bipolar transistor) configuration. The HR/HF series incorporates more than 330 improvements of significance since the original 495-B was introduced. The 495-Series shovels range from 1,250 to 1,482 tons in operating weight.

495HF. The 495HF (High Flotation) is identical to the 495HR except for a widespread undercarriage to accommodate wide (up to 140-inch) crawler shoes. This option has found favor in the oil sands open pit mines in Alberta, Canada, where low ground pressure is essential. The oil sands machines operate with 60-yard dippers and load 400-ton capacity trucks in four passes. The 495HR/HF machines sport a different house from the HD machine including a modern "super cab."

595B. The 351-M electric shovel from the former Marion Power Shovel line became the 595B when Bucyrus International, Inc. took over Marion in 1997. The former Marion flagship electric shovel incorporated fully digital DC static power conversion with diagnostic capabilities. The 1,460-ton 595B was also the first Bucyrus shovel to feature planetary drives in all motions. After seven shovels of this type were sold, the capabilities of the similarly sized 495-Series shovels with advanced AC drive superseded the 595B, and no more were built. From 2005 the hyphen in Bucyrus' model designations was omitted.

Chapter 3
Stripping Shovels

Stripping shovels were probably the most spectacular machines ever built by the Bucyrus companies. They were the giants of the earthmoving industry, designed to move vast quantities of earth and rock in surface mines to uncover valuable minerals like coal, ironstone or limestone. Stripping shovels worked in the same way as regular cable shovels, but unlike their smaller counterparts they didn't need to load another vehicle to carry away the material. Their huge proportions gave them a great range, allowing them to dig from a high face, swing around, and dump the material clear of the working area.

As far we know, there are no stripping shovels still operating today. The last to operate was a Bucyrus-Erie 1950-B known as the Silver Spade that worked in Ohio. It was shut down in 2006 and scrapped the following year. Only three stripping shovels have been preserved at interpretive sites, one in the United States and two in Canada.

Bucyrus-Erie was one of only a very small number of manufacturers worldwide who ventured into designing and building stripping shovels, likely because the market was so small and their use was limited. The same reasons for their limited application also apply to their demise. These can be summarized as follows: i) An enormous amount of engineering and design cost went into these machines, even to prepare a quotation, and then very few of each model would be built. ii) Their production capability was usually specified at the time of ordering to match the output requirements of a particular generating station or processing plant without the need for further expansion. iii) The longevity of the machines, some lasting over 40 years, minimized replacement and saturated the market. iv) Because the stripping shovel sat on the pit floor on top of the mineral being uncovered, it had limited flexibility to operate when compared with a dragline on the surface. v) And sitting on the pit floor limited the depth of material being dug to even the largest of stripping shovels. So by the 1980s, most of the "shallow" coal reserves within the limits of stripping shovels had been exhausted.

Notwithstanding the above negative comments, and in geological conditions appropriate to their use, stripping shovels were the most productive and lowest-cost method of moving huge quantities of material. They were responsible for keeping electric energy costs low for several decades following their inception.

The Bucyrus Company built its first stripping shovel to the designs of the Vulcan Steam Shovel Company which it purchased in 1910. Then, spurred on by competitive stripping shovels marketed by Marion, Bucyrus designed and built its own machines, the first of which, a 150-B went into coal stripping near Pittsburg, Kansas. Throughout the 1910s Bucyrus gained valuable experience from the dozens of stripping shovels it placed in the field. Gradually machines increased in size as overburden became deeper. The giant steam-powered monsters of the 1920s gave way to the more sophisticated models of the 1930s, equipped with electric power and hydraulic leveling. World size records were broken in the 1930s and 1940s as increases in size brought economies of scale and the machines became more reliable.

The "super shovel" era began in the mid 1950s with Marion's 5760 "Mountaineer" and Bucyrus-Erie's 1650-B "River Queen" both in the 60-yard class. A vigorous race ensued between the two major players. It lasted about 10 years and resulted in bigger and bigger machines. The race ended with the world's heaviest machine to move on land, the 15,000-ton Marion 6360 stripping shovel being put into production at the Captain Mine in Illinois. Discounting some long-boom two-crawler shovels built in more recent times, Bucyrus-Erie commissioned its final stripping shovel in 1967.

Thirty years later when Bucyrus International, Inc. purchased Marion Power Shovel in 1997, the two companies that had dominated the stripping shovel market in the United States and worldwide came together after decades of intense rivalry. But no more stripping shovels would be built.

Vulcan Class K. The Bucyrus Company built its first stripping shovel soon after it purchased the Vulcan Steam Shovel Company in 1910. Vulcan had already gained valuable experience in building several part-swing shovels for stripping service with long booms such as this Class K railroad type. Bucyrus then went on to build three fully revolving stripping shovels to Vulcan designs known as Class 5s. They were used in the Kansas coalfield to successfully strip 27 feet of overburden.

54-B. Since its inception in 1939, the 54-B was available as a "high lift" stripping shovel. With 45-foot boom the dipper was rated at 2 cubic yards, correspondingly reduced from the standard 2 1/2 cubic yards on 26-foot boom. The normal twin-stick, rack-and-pinion crowd was replaced by a single-stick rope crowd on the stripping shovel. This employed an auxiliary chain-driven shaft mounted on the boom foot carrying dual drums to operate the twin-rope crowd. The dipper was steadied by two 2-part hoist ropes from a special twin-grooved hoist drum on the machinery deck. In stripping shovel guise the 54-B reached over 100 tons in weight.

88-B. The famous 88-B diesel shovel was also available in "high lift" stripping shovel form in all its series updates from Series I through Series IV until the late 1960s. All carried 47-foot booms, compared with 28-foot for the standard shovel, and dipper capacities were correspondingly reduced by one cubic yard compared with the standard shovel. The air-controlled 88-B Series IV with 300-horsepower Cummins V1710 engine weighed 163 tons.

100-RB. Several 100-RB shovels built in England were ordered as long-boom stripping shovels. They were mainly used in clay pits providing clay for brick manufacturing where it was necessary to excavate a blended product from the full height of a working face. The long-boom shovels usually worked with a 2-cubic-yard dipper, diminished from the normal 100-RB standard 3 1/2-yard dipper.

150-B Steam. When Bucyrus got serious about stripping shovels spurred on by competitive machines from Marion, it designed and built two fully revolving machines in 1911. The rail-mounted steam-powered 150-B carried a 2 1/2-yard dipper on a 60-foot boom and weighed 150 tons in operation. Small by today's standards, the 150-B was nevertheless able to successfully uncover coal seams up to 31 feet deep in the Kansas coalfield where this model was used extensively.

175-B. This was the second of the two stripping shovels released by Bucyrus in 1911. The 214-ton 175-B was offered with steam or electric power, and mounted on crawlers, rail trucks or on skids and rollers. It could remove and dispose of overburden up to 38 feet deep with a 3 1/2-yard dipper on an 80-foot boom. A 175-B owner at Colstrip, Montana, claimed his machine loaded five million cubic yards of lignite in three years without a serious breakdown.

190-B. The 190-B stripping shovel version was offered with booms up to 70 feet long. The one pictured was purchased by Erie Mining Company in 1955 and worked with a boom 56 feet 6 inches long and a dipper holding 6 cubic yards. Operating weight was just under 300 tons. The Bucyrus-Erie two-part boom tied at its midpoint to the A-frame eliminated bending stress in the boom and enabled it to be built lighter, a distinct advantage in reducing inertia when swinging these long booms.

200-B. The 200-B, introduced in 1927, was the smallest of the "modern era" stripping shovels. It offered steam or electric power with Ward-Leonard control, and could be mounted on four crawler assemblies or four rail trucks. The electric-powered crawler version weighed 372 tons when equipped with a 75-foot boom and a 3 1/2-yard standard dipper. About thirteen 200-B strippers built from 1927 to 1943 found their way into gypsum quarries, tin and coal mines.

200-B. This 200-B started life as a rail-mounted coal-loading shovel at Colstrip, Montana, in 1929. It moved to Alberta in the early 1950s and was converted to a crawler machine. The crawler belts were simply wrapped around the original rail flanged wheels and a steering device added. The machine is now preserved at the Reynolds Alberta Museum at Wetaskiwin, Alberta, one of only three stripping shovels preserved in the world.

225-B. When the method of surface coal mining with stripping shovels was proven, Bucyrus designed and built a new monster shovel, the largest yet built, in 1914. The steam-powered 225-B weighed 336 tons and carried a 6-yard dipper on 80 feet of boom. By 1923, over 90 of these huge machines had been erected, making this the no.1 best selling stripping shovel of all time. When Ward-Leonard electric control became available in 1919, the 225-B was one of the first to be so equipped.

270-B. The 270-B electric shovel was introduced in 1960 as the industry's first static-controlled excavator. It incorporated silicon-controlled rectifiers to vary intensity and direction of the main generator excitation current. The first two of this innovative design were sold as stripping shovels with 100-foot long booms and 8-yard dippers. The one shown here operated at Peabody Coal Company's River Queen Mine in Kentucky. Compared to the standard shovel, the long-boom version required more counterweight, longer and wider undercarriage, and 72-inch shoes. Operating weight was 525 tons.

280-B. The 280-B stripping shovel superseded the 270-B in 1962. With increased dimensions over the former model, the 280-B stripping shovel tipped the scales at 615 tons. Electrical power included one 750-horsepower hoist motor, two 188-horsepower swing motors and a 188-horsepower crowd motor. The machine here is one of two 280-B shovels with 80-foot booms at work in a chalk quarry in England. With a 90-foot boom, the 280-B could manage a 9-yard dipper.

320-B. In 1924 Bucyrus built its largest machine to date, the massive 320-B steam-powered stripping shovel carrying a dipper of 8 cubic yards on a 90-foot boom. Early examples were rail-mounted, and these required a sizable ground crew to position the rail sections and keep the shovel literally "on track." Crawler tracks became popular once their mobility was realized, and later versions were electrically powered. The 320-B reigned supreme in the stripping shovel world of the 1920s.

375-B. The crawler-mounted 375-B exhibited the same digging range and capacity as the 320-B, but was heavier built with a more-advanced design. It tipped the scales at just over 600 tons. The Ward-Leonard electric shovel ran with two 150-horsepower hoist motors, a 100-horsepower swing motor and a 100-horsepower crowd motor. One 375-B shovel was shipped in 1935 to a forerunner of Consolidation Coal Company in North Dakota. The other went to Russia in 1940.

385-B. In 1929 the Ward-Leonard electric 385-B stripping shovel at 634 tons filled the gap between the 320-B and the large 750-B. It swung a 12-yard dipper on an 80-foot boom. As with all Bucyrus stripping shovels designed in the 1920s, power for the crawler tracks was transmitted from the main machinery down through a central vertical propelling shaft. At its lower end the drive was split through a series of shafts and gears to power all four crawler assemblies. A two-speed gear drive on each side provided steering.

395-B. Two of the modern AC-drive 395-Bs shovels were equipped with booms 95 feet long compared with a standard boom of 56 feet. Corresponding dippers were reduced to 24 cubic yards but extra counterweight upped the machine weight to 1,079 tons. The second of these two-crawler stripping shovels was shipped to Drummond Coal Company in Alabama in 1988 and represented the last stripping shovel Bucyrus-Erie built.

550-B. The mid-sized 550-B appeared in 1936 and was actually a smaller version of the 950-B stripping shovel released a year earlier. It retained all the features of the larger machine including the two-part boom, tubular dipper arm, and counterbalanced hoist. The 550-B worked with a number of different dipper/boom combinations. A 95-foot boom supported a 20-yard dipper and the machine weighed 880 tons. Only eight of this model were built from 1936 to 1954.

750-B. In 1927 the Bucyrus Company introduced the 12- to 16-yard 750-B, its largest shovel to date. Mounted on four dual crawler assemblies, the 922-ton stripper operated with the Ward-Leonard electric control system consisting of an AC motor generator set supplying DC current to twin hoist motors totaling 500 horsepower, plus two swing motors and a crowd motor all of 100 horsepower.

750-B. Ten of the 12-yard 750-Bs were delivered up to 1930. Then, following a request from the Michigan Limestone Company, Bucyrus-Erie redesigned this shovel employing a counterbalanced hoist system making it more efficient. Dipper size increased 50% to 18 cubic yards but the increase in power consumption was minimal. A moving counterweight at the rear of the house, connected to the dipper hoist drum by ropes, balanced the weight of the empty dipper. Thus all the hoisting power was applied to filling the dipper and hoisting the load.

750-B. In common with other 1920s stripping shovels, two of the four crawler assemblies were attached to a pivoting beam to equalize stresses when passing over uneven ground. To stabilize the machine while digging, a motorized screw jack at each end of the beam locked it tightly to the lower frame. The illustration shows one of the two screw jacks at the far left and the other two crawler assemblies rigidly mounted to the lower frame.

950-B. In 1935 Bucyrus-Erie introduced the 30-yard 950-B as the world's largest shovel. The 1,250 ton stripping shovel pioneered many advanced features on a Bucyrus machine: a two-part pinned boom, a rope crowd with single tubular dipper eliminating torsion stresses, crowd machinery on the revolving frame minimizing swing inertia, independent motors in each of the four crawler assemblies eliminating complicated gear and shaft drive from the upper works, and an automatic hydraulic leveling system that dispensed with the earlier screw-types.

1050-B. Launched in 1941, the 1050-B stripping shovel was an expanded version of the 950-B. Its design was identical in all respects including the counterbalanced hoist system brought forward from the 750-B. The 1050-B boom measured 113 feet long and carried dippers ranging from 26 to 33 cubic yards. A total of twelve 1050-Bs were shipped until the last, a 45-yard machine weighing 1,535 tons, was delivered to United Electric Coal Company's Banner Mine, Illinois, in 1960.

1650-B River Queen. For Bucyrus-Erie, the 1650-B shipped in 1956 marked the start of the "super stripper" era when coal operators could efficiently move into higher stripping ratios and deeper overburden. The 1650-B was designed on similar lines to the smaller 1050-B including counterbalanced hoist and independent motor-driven crawler assemblies. The first, operating a 55-yard dipper on a 145-foot boom, was purchased by Peabody Coal Company for its River Queen Mine in Kentucky, where the machine was named after the mine.

1650-B. After the first 1650-B was shipped, four more were built up to 1964. The others operated with shorter 135-foot booms and dug with dippers up to 70 cubic yards, giving them a dumping radius of 135 feet and maximum dumping height of 96 feet. The 1650-B embodied advanced alloy steels to increase the strength-to-weight ratio. Weights ranged from 2,425 tons to 2,750 tons. *Eric C. Orlemann collection*

1850-B Brutus. Only one 1850-B was built and, although no longer working, it can be seen preserved at West Mineral, Kansas, where it stripped coal from 1963 to 1974. Christened "Brutus," the 1850-B swung a 150-foot boom with a 90-yard dipper. Smaller than the giant, similarly designed 3850-B, it still weighed some 5,000 tons. The advanced electrical design eliminated the need for the counterbalanced hoist of Bucyrus-Erie's earlier shovels.

1950-B Silver Spade. Two 1950-Bs were built. The first, known as the Silver Spade, started work in 1965 at Consolidation Coal Company's Georgetown operations in Ohio. It was one of the most famous stripping machines, not because it broke any size records, but because it was the last to operate—continuing to dig until 2006. It was scrapped the following year. The Silver Spade worked with a 105-yard dipper and 200-foot boom, and its approximate weight in operation was 7,200 tons.

1950-B. Both 1950-Bs employed the "knee-action" crowd, the only Bucyrus-Erie shovels to use this Marion Power Shovel-patented design, and for this a special legal arrangement was formulated. In this design the dipper handle is pivoted to a moveable stiff leg rather than to the boom. During its working life the Silver Spade moved 607,226,370 cubic yards, more than double the volume moved in the Panama Canal excavations.

1950-B Gem. The second 1950-B, named the Gem of Egypt, was commissioned by the same customer in 1967. The 6,850-ton Gem operated with a 170-foot boom and 130-yard dipper, and it turned out to be Bucyrus-Erie's last stripping shovel. Ward-Leonard DC electric power of both 1950-Bs came from eight 1,000-horsepower hoist motors, four 1,000-horsepower swing motors, two 750-horsepower crowd motors, and eight 200-horsepower propel motors. The Gem moved 548,222,376 cubic yards up to 1988, and was scrapped in 1992.

3850-B Big Hog. It was big news when Bucyrus-Erie announced the 3850-B, over twice the size of any shovel built, and the largest mobile land machine of any type up to that time. The 9,000-ton 3850-B, christened the "Big Hog," went to work in 1962 at Peabody's Sinclair Mine in Kentucky. The 210-foot boom provided an operating radius of 210 feet for the 115-yard dipper, designed to move three million cubic yards of overburden per month from a face up to 100 feet high.

3850-B River King. Two years after the first, a second 3850-B began operating at Peabody's River King Mine in Illinois. With a boom ten feet shorter than the first machine (200 feet), the dipper capacity was increased to 140 cubic yards, still the largest ever built by Bucyrus. If scales could weigh this monster they would read 9,350 tons. The River King 3850-B was designed to move 3.5 million cubic yards of overburden per month. When scrapped in 1993, it had moved a total of 731 million cubic yards.

3850-B. The pickup truck and mine visitors demonstrate the gigantic proportions of the 140-yard dipper and its twin bail sheaves. Vertical hydraulic leveling cylinders, each 54 inches in diameter, supported the 3850-B on its four crawler assemblies. A personnel elevator ran up the center of the machine from ground to roof. DC electric motors, eight for hoisting, six for swinging, two for crowding, and eight for propelling, totaled 20,000 horsepower at 460 volts.

Chapter 4
Walking Draglines

In 1911 after purchasing dragline-manufacturing rights from Heyworth-Newman, a Chicago drainage contractor, the Bucyrus Company began building non-walking draglines mounted on rails, skids and rollers, or crawler tracks. They served their owners well, and some large examples were built throughout the 1920s. (See Chapter 5.) But in 1913 Oscar Martinson, an engineer with competitor Monighan Machine Company, also of Chicago, invented the first "walking" device for a dragline.

The simple device consisted of two shoes, one on each side of the machine, hanging by chains to trunnions fixed to a long horizontal shaft running across the machine. As the shaft and trunnions rotated, the leading edge of the dragline's circular tub was raised off the ground and pulled ahead the distance of one step, then lowered gently back to the ground as the shaft continued rotation. The advantages were obvious: lower ground pressure afforded by the circular tub, and steering in any direction by simply pointing the rear of the machine in the desired direction with shoes raised off the ground. Also simple machinery in the upper works with no drive to the base or tub meant lower manufacturing and maintenance costs. In 1925, Martinson improved his earlier design with an improved system replacing the shoe suspension chains by a cam wheel running in an oval track in a frame pivoted to the shoes. This substantially more rigid system would become the standard walking device on draglines for decades to come, and it gave Monighan draglines a distinct advantage over their competition.

During the 1920s, the Bucyrus Company observed Monighan's walking dragline activity with interest. Then in 1931 Bucyrus-Erie Company took a controlling interest in the Monighan company and changed its name to the Bucyrus-Monighan Company in 1934 after taking full control. The former Monighan plant in Chicago continued to build walking draglines under the Bucyrus-Monighan name until 1946, and thereafter under the Bucyrus-Erie name until the plant's closure in 1958 when all dragline manufacture was transferred to Bucyrus-Erie's main plant at South Milwaukee.

From the 1930s to the 1940s many new walking dragline models were introduced, some capturing a world record for size at their introductions. The smaller sizes had optional diesel or electric power, while the larger sizes were powered by multiple electric motors. Starting in the mid 1950s dragline sizes rapidly increased to keep pace with the demand from coal companies who were faced with ever increasing overburden depths. During the 1960s Bucyrus-Erie and its main competitor Marion Power Shovel Company vied to produce the world's largest machine, and both companies succeeded at different times. However, Bucyrus-Erie took the crown in 1969 when the world's largest excavating machine, the 14,500-ton 4250-W walking dragline, better known as Big Muskie, went to work swinging a 220-yard bucket.

The larger draglines were usually ordered with boom and bucket configuration for a specific duty in a specific mine, and while designed from basic standard models, virtually none were identical. Apart from individual design quirks, each new dragline boasted the latest technology. This resulted in later versions of a particular machine being quite different in design and capacity from earlier examples carrying the same model number.

The Arab Oil Embargo of the 1970s resulted in a very busy decade for the walking dragline industry as coal companies ordered machines at an unprecedented rate. While no new physical size records were set, sophisticated designs and computer technology brought increased efficiency and reliability. There were many examples of smaller capacity more efficient machines out-producing larger machines when calculated over a 12-month period. New smaller draglines of modular design also made an appearance in this period. With their fast erection time and easily moved components, they were ideal for short-term coal contracts.

The Ruston-Bucyrus factory at Lincoln, England, also took part in walking dragline manufacture. Beginning in 1939 the 3-W and 5-W models were manufactured over an extended period. More recent models 380-W, 480-W and 1260-W were also made, some exported to customers in the United States during times when the South Milwaukee plant was filled to capacity.

Large draglines returned to the scene in the 1990s when Bucyrus shipped two 2570-WS walking draglines in the 160-cubic yard class: one to the Black Thunder Mine in Wyoming, and the other to the Peak Downs Mine in Australia. Additional draglines of this size are under shipment or erection at the time of writing.

Bucyrus' acquisition of Marion Power Shovel Company in 1997 reduced the number of manufacturers supplying the world with walking draglines to only two, plus some in Russia. The takeover allowed Bucyrus to combine the best design features and patents from the two leading dragline manufacturers. Certain former Marion machines were retained in the product line and are now sold as Bucyrus machines with updated technology. Bucyrus' only other competitor, P&H Harnischfeger, has not built a dragline with a capacity greater than 115 cubic yards at the time of writing, leaving the large dragline field open with no competition for Bucyrus.

1-T. After experimenting with several prototype machines Oscar Martinson of the Monighan Machine Company applied his "Martinson Tractor" walking device to the first commercial walking dragline. The 1-T with one-yard capacity on a 40-foot boom weighed 29 tons, and was gasoline-powered with Monighan's own 55-horsepower 3-cylinder engine. It was produced up to 1926 when 117 had been sold. The slightly larger **1 1/2-T** model with 1 1/2-yard bucket on a 50-foot boom was added to the line in 1922.

2-T. The 2-T, launched in 1915, carried a standard bucket of 2 cubic yards on booms up to 70 feet long. The Martinson walking device revolutionized the dragline industry, freeing the machines from their restrictive rails or skids and rollers, and lowering the ground pressure under the large circular base. The 2-T was offered with steam or kerosene power, the latter equipped with a 70-horsepower 2-cylinder engine with twin flywheels. Monighan shipped a single model **2 1/2-T** in 1922 carrying a 2 1/2-yard bucket on a 60-foot boom.

3-T. The 3-yard 3-T was introduced almost simultaneously with the 1-T in 1913. It carried a standard 60-foot boom, and was offered with either steam power or 90-horsepower kerosene engine, and was produced up to 1925. An identical machine, the 3-S, was built with skid and roller mounting. (See chapter 5 for description.) Three walking draglines designated model **3 1/2-T** were shipped in 1922 with 3 1/2-yard buckets. Rounding out the T-series was the 4-yard **4-T**, six of which were built from 1924 to 1925.

2-W. The 2-W with a 2-yard bucket on a 60-foot boom appeared in 1926. It embodied improved features making it easier to move from job to job with minimal dismantling required. Its circular base measuring 14 feet in diameter resulted in a low ground pressure of 6.4 psi. The 2-W power unit was a 2-cycle, 2-cylinder diesel engine developing 80 horsepower.

3-W. The 3-W was the first machine to employ Monighan's improved walking system introduced in 1925. It was fitted with a similar type of 2-cylinder diesel engine to the 2-W, but larger at 120-horsepower. Most of these models were sold with 70-foot booms and 3-yard buckets, but longer booms up to 100 feet were supplied with 2-yard buckets. Operating weight averaged 115 tons.

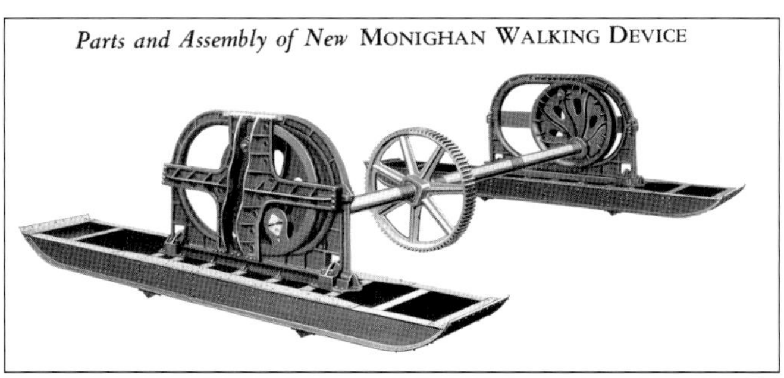

Monighan Walking System. In 1925, Martinson improved the Monighan dragline walking device by replacing the shoe suspension chains with cam wheels running in an oval track. The walking shaft running across the machine that drove these wheels was driven from the main machinery through a jaw clutch. Bucyrus-Erie later employed this simple but very successful system on draglines up to 70-cubic-yards capacity.

1-W. Monighan's improved walking system resulted in the former T-series draglines being immediately obsolete, and a new range of machines identified as the W-series took over. The 1-yard 1-W with 40-foot boom was introduced in 1926. It was the smallest in the W-series and available with diesel, gasoline or electric power. Operating weight was 32 tons. Two slightly larger machines designated **1 1/2-W** with 60-foot booms and 1 1/2-yard buckets were shipped the same year.

3-W. Fifteen of the first-series 3-Ws were sold up to 1930. Then after a lapse of nine years, the 3-W was re-introduced by Bucyrus-Monighan in 1939 as a new machine of modern design. Range and capacity were the same as the former model but weight was reduced to 106 tons. This machine was also built by Ruston-Bucyrus in England beginning in 1941. The 3-W power options were either straight electric with Ward-Leonard control, or a 190-horsepower Fairbanks-Morse 5-cylinder diesel engine driving the main drums mechanically and a generator for the electric swing drive.

4-W. The 4-W appeared in 1926 as one of the first to boast the improved Martinson walking mechanism. Like the 1-W, 2-W and early 3-W designs, the hoist, drag and swing motions were operated through friction clutches driven by the diesel engine, which on the 4-W was a 3-cylinder Fairbanks-Morse rated at 180 horsepower. The 232-ton 4-W was supplied with booms ranging from 80 feet with a 5-yard bucket, to 120 feet with a 3-yard bucket.

5-W. After Bucyrus-Erie took over Monighan to form Bucyrus-Monighan, the first new dragline to appear from the Chicago plant was the 5-W in 1934. This model turned out to be one of the most successful small walking draglines ever built. It swung a 5-cubic-yard bucket on a 120-foot boom or a 4-yard bucket on a 135-foot boom. Electric or diesel power was available and the early diesel models furnished with a 250-horsepower "marine-type" 5-cylinder engine. Later models ran a 300-horsepower Fairbanks-Morse 6-cylinder engine.

5-W. The picture shows one of the 5-Ws built by Ruston-Bucyrus in England over a production life that lasted until 1971. The machine was the same as the American-built model except its diesel engine was a 5-cylinder Ruston developing 320 horsepower, and later models included a mast to better support longer booms. The 5-W still holds the all-time best-selling record for a walking dragline, with 79 built in the Chicago plant and a further 62 built at Lincoln, England.

6-W. The 6-yard, 316-ton 6-W was another of the early walkers released in 1926. In common with most other Monighan-designed draglines, the 6-W employed a belt drive from the engine to the main draw works. On the 6-W and larger, an electric swing drive was provided with Ward-Leonard control instead of the two asbestos-lined cone clutches on the smaller machines. Of the 24 6-Ws built, no less than 12 were shipped to the U.K. to assist with coal stripping after World War II.

7-W. Launched in 1942, the 7-W was the last Monighan-designed dragline introduced under the Bucyrus-Monighan name. After 1946 it continued as a Bucyrus-Erie machine and remained in production until 1971. With an operating weight of 346 tons, the 7-W worked with booms from 140 to 180 feet and respective bucket sizes from 7 to 5 cubic yards. Ward-Leonard electric or diesel power was available, the latter fitted with a Cooper-Bessemer GS-6 engine rated at 435 horsepower at 525 rpm.

9-W. The popular 9-W first appeared in 1938 and offered booms from 160 to 200 feet long with corresponding buckets of 10- to 7-cubic-yards capacity. The electric version came with a 450-horsepwer AC synchronous motor-generator set and Ward-Leonard DC motor control. The diesel version was provided with a 6-cylinder Fairbanks-Morse engine having a 12-inch bore and 15-inch stroke developing 630 horsepower at 360 rpm. Operating weight was 511 tons.

10-W. The 10-W offered long boom capability up to 200 feet, with a 10-yard bucket. The first, introduced in 1934, was diesel powered with a 450-horsepower engine running at a low 260 rpm. A wide belt drive connected the engine to the draw works and a generator for the electric swing. The remaining four 10-Ws shipped up to 1939 were straight electric Ward-Leonard-controlled machines with two 150-horsepower DC hoist motors and two 100-horsepower DC swing motors.

15-W. Only three 15-Ws were built, all shipped in 1940. Operating with booms up to 215 feet long and 13-yard buckets, the 15-W was one of the largest draglines with hoist and drag drums controlled by clutches and brakes as opposed to close-coupled electric motors. The necessarily large diameter outside band type brakes and clutches were asbestos lined; the clutches were actuated by hydraulic cylinders. The 15-W tipped the scales at 846 tons.

6150. This nomenclature, used by Bucyrus-Monighan for a time during the 1930s, was an attempt to indicate the capacity of the particular machine. The 6150 signified a 6-yard bucket on a 150-foot boom, but only the first machine shipped actually possessed this capacity. Subsequent machines swung booms up to 170 feet. A 300-horsepower Fairbanks-Morse diesel powered the 6150 which had a similar machinery layout to the 6-W.

6160. The 6160 superseded the 6150 in 1932 with slightly increased ratings to permit a 6-yard bucket on a 175-foot boom, or 5-yard bucket on a 200-foot boom. Power was boosted to 350 horsepower and weight increased to 453 tons, some 50 tons heavier than the 6150. The 6160 was also available with a special 36-foot diameter tub, six feet greater than standard, reducing ground bearing pressure to 6.6 psi.

8160. The 8160 was a diesel-driven dragline powered by a 5-cylinder Fairbanks-Morse engine having a 14-inch bore and 17-inch stroke and producing 450 horsepower at 260 rpm. The 8160 design followed that of the 6150 closely, but with increased capacity for 8- or 10-yard buckets possible on a 160-foot boom. Only three 8160s were shipped, all in 1931. The last survivor, located in Oklahoma, was re-engined in 1969 with a 550-horsepower Caterpillar D379.

180-W. Making an appearance in 1954, the 180-W was an updated version of the American-built 5-W walking dragline with similar range and capacity, but with increased power and weight for faster cycle times. A V12 Caterpillar D397 engine rated at 500 horsepower powered the diesel machine, while the electric version housed a Ward-Leonard electric drive system with AC motor generator sets. The 180-W also boasted a new roomier, rectangularly shaped house.

200-W. From 1945, dragline requirements in the 4- to 6-cubic-yard class were met by the 200-W, and 55 sold up to 1955. The machine was offered with boom lengths up to 140 feet and diesel or electric power could be specified, the latter provided by a Cooper Bessemer GS-6 engine of 385 horsepower. The versatile 200-W was used in many diverse applications and shipped to countries around the globe including Pakistan, Thailand, Australia, and Sweden.

380-W. In 1977 Bucyrus-Erie announced the 380-W, a totally new and revolutionary dragline design. Built in modular units bolted together, the 500-ton dragline was designed to be erected or moved quickly in just a few weeks instead of months with a conventional machine. The concept was ideal for contractors working small pockets of coal in the 1970s coal boom. The 380-W ranged in the 10- to 16-cubic-yard class with booms to 200 feet. Of the twenty-four 380-Ws shipped, six were manufactured by Ruston-Bucyrus in England.

450-W. Introduced in 1948, the 450-W was a large diesel-powered dragline in the 10-yard class. An 8-cylinder Cooper-Bessemer JS-8 provided 850 horsepower at 450 rpm to the main draw works through a spur gear train to the hoist and drag drums. The drums were driven and controlled through air-actuated asbestos-lined friction clutches and brakes. The engine also drove a DC generator providing Ward-Leonard electric power for two swing motors.

480-W. In 1955 Bucyrus-Erie replaced the 450-W with the similarly designed 480-W, but with diesel or electric power options. The diesel version used the same engine as the former model but uprated to 1,000 horsepower. The popular 480-W initially handled 12 yards on a 175-foot boom, but over its long production run capacity crept up to 18 yards and weight increased from approximately 700 tons to 800. The last machines were shipped in 1979.

500-W. Introduced in 1946, the 500-W was a fully electric machine and actually smaller in capacity than the 480-W. Its buckets ranged from 8 to 10 cubic yards on corresponding booms from 205 to 165 feet long, and specified operating weight was 650 tons. The neatly-designed 500-W distinguished itself from the 450-W and 480-W with its independent DC hoist and drag motors geared to each drum, as opposed to a single power source driving both drums through clutches and brakes.

650-B. Another dragline first shipped in 1946 was the 810-ton 650-B. The fully electric machine was furnished with booms from 175 to 215 feet and buckets from 16 to 12 cubic yards respectively. Like the smaller 500-W, the 650-B's digging motions were driven by independent DC electric motors powered through AC motor-generator sets with Ward-Leonard control. Two hoist and two drag motors of 250 horsepower each, and three swing motors of 75 horsepower were included.

680-W. Moving forward from the 380-W, Bucyrus-Erie designed and built this larger modular walking dragline in the 26-yard class. The 680-W, appearing in 1988, was a "clean-sheet" design featuring state-of-the-art "Acutrol" controlled-frequency AC electric drive with static control, and a unique machinery layout. An "inter-lock" clutch allowed hoist and drag drums to lock together so that as the bucket was paid out, pull on the drag rope assisted the hoist motion, thereby saving energy.

770-B. Introduced in 1954 the 20-yard 770-B was essentially a smaller version of the 1150-B with similar appearance. The straight electric machine embodied AC motor-generator sets powering Ward-Leonard-controlled DC motors for all functions: two 500-horsepower hoist motors, two 500-horsepower drag motors, three 188-horsepower swing motors, and two 150-horsepower walk motors. Machines were sold with boom lengths of 195 feet or 235 feet and operating weights around 1,000 tons.

800-W. Only two 800-Ws were shipped, one in 1966, the other in 1970. Sized at 28 cubic yards, they sported Bucyrus-Erie's modern curved house design, first appearing on the 1450-W in 1963. The 800-W also featured Bucyrus-Erie's innovative triangular boom with tubular members and pressurized main chords. Loss in pressure would signal a structural crack. With a 195-foot boom, the 800-W weighed 1,130 tons in working order.

950-B. Soon after Bucyrus-Erie acquired the Monighan manufacturing rights, it built the world's largest dragline, a single 950-B weighing just over 1,000 tons. Shipped to Brazil in 1935 for stripping overburden in a large limestone quarry, it swung a 12-yard bucket on a 250-foot boom, the world's longest at that time, and sat on a 45-foot diameter tub. Regarded as an engineering masterpiece, the 950-B laid the foundation for much larger Bucyrus-Erie walking draglines over the next three decades.

1150-B. Another milestone machine was the 1150-B launched in 1944. Developed from the earlier 950-B walker, the 1150-B carried buckets in the range from 20 to 25 cubic yards with booms from 180 to 215 feet, and weighed approximately 1,200 tons. Sixteen 1150-Bs were delivered in the United States prior to 1950. After only a few years' work, four of these were relocated to the United Kingdom to assist with coal stripping. All four worked well into the 1980s, and one is now preserved in Yorkshire.

1250-B. The 1250-B was an upgraded version of the 1150-B which it replaced in 1951. Of similar design and appearance, the 1250-B's increased working range allowed operation with a 250-foot boom and 24-yard bucket, or 200-foot boom with 30-yard bucket. Its 50-foot diameter tub and working weight of 1,465 tons were up from 45 feet and 1,250 tons in the 1150-B. Motor power was the same, but a fourth swing motor was added as opposed to three installed in the former machine.

1250-W. Replacing the 1250-B in 1959, the 1250-W possessed probably one of the most interesting booms installed on a dragline. Built in two parts hinged together, and a mast pivoted on the lower section, the boom was made with lightweight alloy steel to support additional capacity. The motors remained the same size as on the 1250-B, i.e. two 1,000-horsepower motors each for hoist and drag, four 150-horsepower swing motors, and two 300-horsepower walk motors.

1260-W. Thirty-three 1260-W walking draglines were shipped from 1965 to 1990. Most were sold with buckets ranging from 30 to 40 cubic yards and corresponding booms from 285 to 225 feet. The first nine machines utilized a similar house to the 1250-W, but machines after that sported the streamlined curved house design first introduced on the 1450-W. The 1260-W also carried Bucyrus-Erie's patented triangular boom with gas-filled chords

1260-W. Six of the 1260-Ws were built by Ruston-Bucyrus at Lincoln, England. This one swings a 32-yard bucket on a 260-foot boom and shows the later house design. Notice that the triangular boom on this machine is mounted with the apex at the top. Others were inverted with the apex at the bottom as in the previous illustration. Average operating weight of the 1260-W, depending on boom length, was 1,845 tons.

1300-W. The 1300-W first shipped in 1971 was a scaled down version of the 1370-W released two years earlier. It normally carried buckets from 40 to 45 yards, but some long-range machines were shipped with 325-foot booms and buckets down to 28 yards. Operating weight was 2,250 tons. The triangular booms on the 1260-W and 1300-W could be mounted with the apex on top or on the bottom. The picture shows "Nessie" stripping coal in Alberta, Canada.

1350-W. The 1350-W was the fist of the 1300-series walking draglines launched in 1967. Its boom range from 285 to 320 feet permitted buckets from 40 to 60 cubic yards, depending on the material. Independent electric motor drive with Ward-Leonard control comprised four 750-horsepwer motors for hoist and drag, four 500-horsepower swing motors, and two 500-horsepower walk motors. A modern streamlined house and triangular boom identified this model.

1360-W. Nudging up in size was the 1360-W, first shipped in 1975. With increased power from its four 1,024-horsepower hoist motors, four 800-horsepower drag motors, and operating weight upped to 3,116 tons, the 1360-W was able to work with a 50-yard bucket on a 325-foot boom. Machinery layout, triangular boom and overall design followed closely that of other draglines in the 1300-series.

1370-W. Introduced in 1970, the 1370-W became a very popular dragline for overseas customers. Of the 38 units sold up to 1984, only 12 remained in the United States, 23 were shipped to Australia and three to Africa. With an average weight of 3,230 tons, booms ranged from 267 to 325 feet long with corresponding buckets ranging from 55 to 62 cubic yards.

1500-W. The only two 1500-Ws built were shipped to Texas Utilities in 1970/71. They were equipped with 285-foot booms, 70-yard buckets and their specification showed an operating weight of 2,900 tons. Although their range and suspended load capability was slightly more than the earlier 1550-W model they superseded, their weight was some 300 tons lighter, reflecting a more efficient and modern design.

1450-W. All seven of the 1450-Ws were equipped with 60-yard buckets on 250-foot booms. When the first started work in 1963 at P&M's Colonial Mine in Kentucky, it was for a brief time the world's largest dragline. It also pioneered a new machinery house incorporating a streamlined curved design. The 1450-W possessed an abundance of Ward-Leonard DC power with four hoist, four drag, five swing and four propel motors totaling 10,375 horsepower. Operating weight was 3,010 tons.

1550-W. Two 1550-Ws were built in 1965, both specified with 65-yard buckets. One worked in Illinois for Consolidation Coal Company, the other, owned by contractor Derek Crouch and named Big Geordie, spent its working life stripping coal in northern England. The 1550-W tub measured 55 feet in diameter and machine weight was 3,200 tons. It boasted sixteen DC motors totaling 14,800 horsepower for hoist, drag, swing and propel.

1570-W. In the 70-yard class, the 1570-W, and the former Marion 8200 retained as a Bucyrus product, captured the mid-size walking dragline market with 80 units sold between them. The first 1570-W, with a 310-foot boom, was sold to Drummond Coal in Alabama in 1973. Other 1570-Ws sold up to 1991 included booms from 285 to 345 feet, buckets up to 80 cubic yards and operating weights to 3,750 tons. They featured the standard Bucyrus-Erie triangular boom.

2550-W. The 2550-W introduced in 1964 swung a bucket of 75 yards on a 275-foot boom and rested on a 65-foot diameter tub. It was the first application of a triangular boom on a large walking dragline. As with all subsequent triangular boom designs, the three main chords were filled with gas under pressure so that leaks indicating cracks could be reported in the operator's cab.

2560-W. Moving a step upwards from the 2550-W, the 2560-W was launched in 1969. Operating weight was some 400 tons more at 4,825 tons, and its 22 DC motors totaled 21,000 horsepower, 4,000 more than the former model. Two 2560-Ws were built, the first with a 295-foot boom and 85-yard bucket went to Midland Coal's Elm Mine, Illinois, the second operated with a 275-foot boom and 90-yard bucket at Peabody's Universal Mine, Indiana.

2570-W. Regarded as Bucyrus-Erie's "bread-and-butter" large dragline, the 2570-W was a popular machine in large surface coalmines with 27 units sold between 1971 and 1983. It carried buckets from 80 to 115 cubic yards and swung booms measuring from 285 to 400 feet. Operating weights averaged 6,200 tons. The first machines utilized the single triangular boom style, then a twin triangular type was adopted. Later models employed a four-chord trapezoidal-type boom.

2570-WS. A "super" version of the Bucyrus-Erie 2570-W dragline, the 2570-WS, bearing little resemblance to the former model, was erected at Arch Coal's Black Thunder Mine, Wyoming, in 1992. With a load capability of 675,000 pounds or 160 cubic yards on a 360-foot boom, this dragline is the largest currently operating in North America. In 2000 another 2570-WS started work for BHP Coal at Peak Downs, Australia. With a maximum suspended load of 800,000 pounds, this dragline holds the title of the world's largest operating dragline today.

3270-W. At the top end of the scale, Bucyrus was successful in placing two huge 3270-W draglines with Amax Coal Company who put them to work in 1979. Designed on modern principles, the 3270-Ws worked at Ayrshire Mine, Indiana, and at Delta mine in southern Illinois. They were the second-largest draglines ever built with 176-cubic-yard buckets, 330-foot booms and operating weight of 8,718 tons.

4250-W. Christened "Big Muskie" this dragline was probably the most famous and most definitely the largest ever built. Holding 220 cubic yards on a 310-foot boom and weighing 550 tons with contents, Big Muskie's bucket was the largest ever mounted on an excavator. When first commissioned the 4250-W's weight was quoted as 13,500 tons, but subsequent major structural additions were estimated to add another 1,000 tons to this figure.

4250-W. When starting work in 1969 the 4250-W dragline was some 50% larger than the previous world's largest. The previous title holder was the Marion 8900 at 145 cubic yards which went to work a couple of years earlier. The 4250-W electric motor arrangement included ten hoist motors, eight drag motors and ten swing motors. The unique propel system, utilizing four giant hydraulic jacks to raise the entire 105-foot diameter tub off the ground, held 20,000 gallons of hydraulic oil.

4250-W bucket. Some perception of the 220-yard bucket size is seen here. The bucket was suspended by four of the biggest wire ropes ever manufactured at that time—measuring five inches in diameter. Hoisting was accomplished through two separate drums eleven feet in diameter, each winding two of the 5-inch ropes. During her life Big Muskie moved 608 million cubic yards of overburden at the Central Ohio Coal Company's Muskingum Mine, Ohio. She was scrapped in 1999.

8200. This former popular Marion model was adopted as a Bucyrus product after the takeover in 1997. The first Bucyrus 8200 was sold to BNI Coal Company in North Dakota and commissioned in 2004. This machine swings a 77-yard bucket on a 355-foot boom and weighs 5,210 tons. Following its proven tradition as a DC drive machine, the Bucyrus version embodies a host of state-of-the-art electronic controls and diagnostics.

8750D. The latest in dragline technology was applied to a new dragline shipped in 2005 to China from South Milwaukee. The dragline, a model 8750D with a 117-cubic yard bucket derived from the former Marion model, is the first large dragline from any manufacturer to be sold into China. It's also the first dragline to be entirely powered with a Siemens AC static drive system, and introduces direct (gearless) drive for the hoist and drag motions.

Chapter 5
Non-Walking Draglines

In 1910 Bucyrus entered the dragline market after purchasing the manufacturing rights of the Heyworth-Newman dragline excavator. James O. Heyworth was a contractor who had built several successful dragline-type machines for use on his own drainage contracts in the Chicago area. At the time of purchase Bucyrus management stated, "The dragline type of machine will be one of vital importance in our line of manufacture and it should make serious inroads upon the steam shovel business."

The following year, Bucyrus introduced its first in-house designed dragline, the Class 24, incorporating many of the Heyworth-Newman patents. The steam-powered dragline was also the largest built up to that time, and carried a 3 1/2-cubic yard bucket on a 100-foot boom.

As noted in Chapter 4, Oscar Martinson of the Monighan Machine Company invented the walking dragline in 1913, but Bucyrus did not have access to this type of equipment until it acquired Monighan in 1932. Until that time Bucyrus developed a full line of Class-series non-walking machines, including a Class 14 introduced in 1912, the first dragline to be mounted on crawler tracks. Steam, gasoline, diesel or electric power was offered.

Many of the early Bucyrus draglines were mounted on rails or skids and rollers, with no propel drive other than the dragline's own bucket to pull itself along. Both methods were very labor-intensive. The rail-mounted machine needed a ground crew to lift and move the rail sections as the machine moved ahead. In the case of the skid and roller machine, an even larger ground crew was employed to place the heavy timbers on which the machine rolled, and manhandle the hardwood rollers from the rear to the front as the machine dragged itself along. All precautions were taken to avoid the machine running away and departing from its intended rail or roller path!

As draglines increased in size, crawler mounting became the preferred choice for machine support and movement. Some large crawler machines were built in the 1920s with similar 8-crawler undercarriages as found on stripping shovels. But their maximum size was limited due to excessive ground pressure under the crawlers, a problem not solved until the walking dragline entered the scene.

For the most part, the early draglines were built as single-purpose machines, i.e. they were not intended to convert to shovels. Thus for a few years Bucyrus developed its shovels and draglines in separate departments until the smaller universal excavators became more popular in the early 1920s. The non-walking single-purpose dragline category died in the early 1930s as small universal excavators took over the smaller sizes and walking draglines took over the larger sizes. Nevertheless over 370 of the Class series non-walking draglines were produced from 1911 to 1937.

Heyworth-Newman. This is the dragline that attracted Bucyrus' attention in 1910 and caused it to enter the dragline market. The steam-powered machine is shown with a 3 1/2-yard patented Newman bucket with a 3-part hoist arrangement, and a 100-foot boom. This Newman bucket was soon superseded by the Page automatic dumping type. Bucyrus developed its first in-house draglines base on the Heyworth-Newman machine.

Class 7. The smallest of the Class series was introduced in 1915 as a 1-yard capacity dragline with a standard boom of 42 feet. Operating weight was 35 tons. Continuing in the Bucyrus product line until 1923, the Class 7 was offered with steam, gasoline or electric power and a variety of undercarriage options. Skid and roller mounting was standard but machines with rail, traction wheels, and either two- or four-crawler undercarriages were built.

Class 9 1/2. The standard Class 9 1/2 dragline worked with a 1 1/2-yard bucket and a 45-foot boom, with steam, gasoline and electric versions available. Mountings offered were either crawler or skids and rollers, with operating weights of 43 tons and 60 tons respectively. The U.S. Reclamation Service purchased 16 Class 9 1/2 draglines for its important drainage projects in Texas and New Mexico.

Class 14. Bucyrus launched the 2-yard Class 14 dragline in 1912, and as expected it resembled the former Heyworth-Newman machine from which it was derived. The standard boom was 70 feet long and operating weight for the skid and roller machine as shown was 63 tons. Steam, gasoline, electric and diesel-electric power options were available.

Class 14 Lower. The Class 14's claim to fame was that it was the first excavator of any type to be fitted with crawler tracks. The illustration shows the prototype assembly utilizing hardwood shoes. These were soon followed by steel castings. The Class 14 crawler version employed four separate assemblies all driven from the upper works. The crawler option, strongly advocated by Bucyrus in its sales literature, increased machine weight to 90 tons.

Class 14. The Class 14 remained in production until 1930 and this illustration represents a later model. It is working for T.J. Clark Construction and is digging a drainage ditch near Estherville, Iowa. A two-cycle, two-cylinder diesel engine directly connected to a generator provided the power to a 50-horsepower swing motor and a 150-horsepower main motor. Hoist and drag motions were operated through clutches and brakes.

Class 20. The year after Bucyrus launched its first dragline, the Class 24 in 1911, the Class 20 appeared as a scaled-down version of the larger machine. Although regularly equipped with an 85-foot boom and a 2 1/2-yard bucket, 75-foot and 95-foot booms were available with 3-yard and 2-yard buckets respectively. The less popular Class 20 was mounted on rails or skids and rollers and steam or electrically powered.

Class 24. The first Bucyrus in-house designed dragline after purchasing the Heyworth-Newman manufacturing rights was launched in 1911. It swung a 3 1/2-yard bucket on a 100-foot standard boom. 85-foot and 115-foot booms were also available with buckets of 4 1/2 and 2 1/2 cubic yards respectively. Later updates to this well-accepted machine allowed a standard 4-yard bucket on a 100-foot boom.

Class 24. This rail-mounted Class 24 gives some idea of the labor intensity needed for non-walking draglines. Initially, the Class 24, mounted on skids and rollers, weighed 144 tons. When the last of the crawler-mounted machines were shipped in 1930, their weight had increased to 246 tons. The Class-series dragline model numbers (up to the Class 24) indicated the diameter of the swing roller circle in feet, indicating a diameter of 24 feet for the Class 24.

Class 24. The last surviving Class 24 is preserved at the Reynolds Alberta Museum, Wetaskiwin, Alberta, Canada. It is a standard steam-powered machine shipped in 1917. The popular Class 24 was built with all power sources available including compressed air. The steam version was fitted with a large horizontal locomotive-type boiler 58 inches in diameter by 16 feet long and according to the specification burned about four tons of coal per 10-hour shift.

Class 24. Celebrating "erection completed" of a late model Class 24 diesel-electric dragline in India, this group poses for the camera. Power for this late-model diesel-electric was provided by a Fairbanks-Morse two-cycle, three-cylinder engine with 14 in. bore and 17 in. stroke, developing 180 horsepower at 257 rpm. The crawlers on one side were connected to an equalizer beam to accommodate irregular ground surfaces. (See under Class 375.)

Class 80. This was the dragline version of the 80-B shovel first built in 1921. Its standard bucket held 2 1/2 cubic yards when swung from a 60-foot boom. Smaller buckets with 70-foot and 80-foot booms were also available. The Class 80 was steam or electrically powered, and was mounted on rails or on four crawler tracks. The two crawlers on one side were connected to an equalizer beam as described for the Class 375.

Class 111. Built from 1931 to 1939 the Class 111 was a special 3-yard diesel dragline designed for levee building or drainage work. It was another of those machines equipped with a low-revving diesel, in this case a 5-cylinder Fairbanks-Morse with 10-inch bore and 12 1/2-inch stroke, and developing 210 horsepower at 425 rpm. Its 22-foot long crawlers with 42-inch shoes gave a low ground pressure of about 11 psi, comparable to that of a walking dragline. It also boasted full air control for all clutches and brakes.

Class 175. This was the dragline version of the 175-B stripping shovel, one of the first such machines made by Bucyrus beginning in 1912. The Class 175 was rail-mounted and offered with steam or diesel-electric power. Later electric versions embodied some of the earliest applications of Ward-Leonard control. The Class 175's standard 4-yard bucket worked in conjunction with a 125-foot boom. Operating weight was 219 tons.

Class 180. The Class 180 special dragline excavator built by Bucyrus from 1915 to 1920 was steam powered. It was mounted on four rail trucks, one at each corner of the lower frame, which ran on standard-gauge railroad tracks. The Class 180 dragline worked with buckets of around 4-cubic-yards capacity, fitting in between the Class 175 and Class 200 machines.

Class 200. This was another adaptation of a stripping shovel for dragline service. The 200-B shovel first appeared in 1927, soon followed by the first Class 200. The machine was either supported on rails, or on an 8-crawler undercarriage in which two pairs of crawlers on one side were connected to an equalizer beam similar to the Class 24. Power options were steam or Ward-Leonard electric. Buckets of 5 or 6 cubic yards could be supplied on corresponding 125-foot or 113-foot booms.

Class 225. The 225-B was the no.1 all-time best seller of stripping shovels with over 90 produced from 1915 to 1923. The Class 225 was the dragline version of the same machine mounted on an identical rail-mounted base. Each rail truck was driven by chains through a series of gears and shafts from the upper works. This 7-yard machine worked in the Florida phosphate fields for Virginia-Carolina Chemical Corporation. *Photo courtesy of Jim Dalton*

Class 230. Although rated at 6 cubic yards, the Class 230 dragline offered a great operating range with booms up to 160 feet supported at a low 18 degrees to give a 171-foot operating radius. The four rail trucks or crawler assemblies were positioned at each corner of a 30-foot square base, making the 330-ton machine exceedingly stable. Steam or diesel-electric power options were available, with the latter requiring two 190-horsepower engines close coupled to generators.

300-D. The 300-D was the crawler version of the 380-W walking dragline. The single machine built was diesel powered with a Caterpillar D349 diesel rated a 1,130 horsepower at 2,000 rpm. Like its walking counterpart the 300-D was modular-designed for rapid moving and erection. The 400-ton 300-D carried a 10-yard bucket, and was mounted on a uniquely designed 4-crawler undercarriage with great flexibility for negotiating uneven ground surfaces.

Class 320. The 320-B was the largest Bucyrus stripping shovel when introduced in 1923. A year later the first Class 320 dragline was shipped to Florida for the Southern Phosphate Company carrying a 6-yard bucket on a 143-foot boom. Later machines worked with 8-yard buckets. The Class 320 was touted as the world's largest and most powerful dragline yet built. Ward-Leonard electric or steam power were available. The latter required an 80-inch diameter locomotive-type boiler 20 feet long.

Class 375. First appearing in 1929, the Class 375 was the largest of Bucyrus' non-walking draglines. The 375-B stripping shovel version was developed two years later after four of the draglines had been sold. The Class 375 was available with steam, electric or diesel-electric power. Bucket range spanned from 5 to 10 cubic yards with corresponding booms from 160 to 125 feet. As with most of the Bucyrus Class-series draglines, rail or crawler mountings were available.

Class 375 Lower. This was the 8-crawler undercarriage of the Bucyrus Class 375 dragline, a design common to most Bucyrus large crawler draglines and early stripping shovels. The two crawler assemblies on one side of the undercarriage were connected to a pivoting beam. This compensated for uneven ground when moving, and eliminated twisting stresses in the frame. When digging, screw jacks at each end of the beam were wound up using the hand wheels to form a rigid, stable base.

Chapter 6
Railroad Shovels

In 1882 the Bucyrus Foundry and Manufacturing Company received an order from the Toledo and Ohio Central Railroad to build a steam shovel. This important milestone was the Company's first entry into manufacturing excavating equipment. The machine delivered was known as the "Thompson Iron Steam Shovel and Derrick," named after its designer John Thompson, Manager of Manufacturing at Bucyrus. Steam shovel orders for the Northern Pacific Railroad and the Savannah, Florida & Western Railroad followed shortly, and all were delivered within a few months.

Bucyrus shovels became known for their robust strength and substantial construction, and soon gained an enviable reputation. As early as 1883, the Bucyrus Thompson Shovel won first prize at the National Exposition of Railway Appliances in Chicago. This early publicity and reports of excellent performance generated further machine sales, with 59 selling by 1889.

These early shovels consisted of a steam boiler, steam engines, draw works, and shovel boom, arm and dipper. All were mounted on a standard-gauge heavy-duty railroad car fitted with outriggers that were set down during digging. The shovels were not fully revolving—the boom, digging arm and dipper swung as a unit at the front of the railroad car. Because this part-swing shovel configuration was the only type available for almost two decades, and because most of the early sales were to railroad companies, this type of shovel became universally known as the Railroad Shovel, even when applied to other applications in later years.

America's involvement in the Panama Canal construction between 1904 and 1914 resulted in a real boost for Bucyrus steam shovels, with 77 supplied out of a total of 102 employed on the job. Such extensive orders for similar products permitted the Company to familiarize itself with the economics of producing machines in large lots. And the substantial publicity the Company gained by its important contribution to the Canal construction has persisted in advertising to the present day. In a 1909 catalog, Bucyrus claimed it had shipped 1,100 steam shovels by that date.

Bucyrus bolstered its position in the railroad shovel business by purchasing two important steam shovel manufacturers in 1910/1911. These were the long-established Vulcan Steam Shovel Company, and the Atlantic Equipment Company, whose substantial shovels were designed by A.W. Robinson, former Chief Engineer at Bucyrus.

Fully revolving excavators, usually of small size, appeared in the early part of the 20th Century but the heavy-duty ruggedness and durability of the railroad shovel remained unchallenged until the mid-1920s. The advent of fully revolving quarry and mine shovels, led by the Bucyrus 120-B in 1925, caused sales of the railroad shovel to rapidly diminish until Bucyrus shipped its last in 1930.

Bucyrus' line of railroad steam shovels, which dominated the company's products for the first four decades if its existence, covered every size from 25 tons to 110 tons in approximately 5-ton increments. Although all the major sizes are covered in this chapter, a few models from each design era had to be omitted because of space limitation or lack of available documentation.

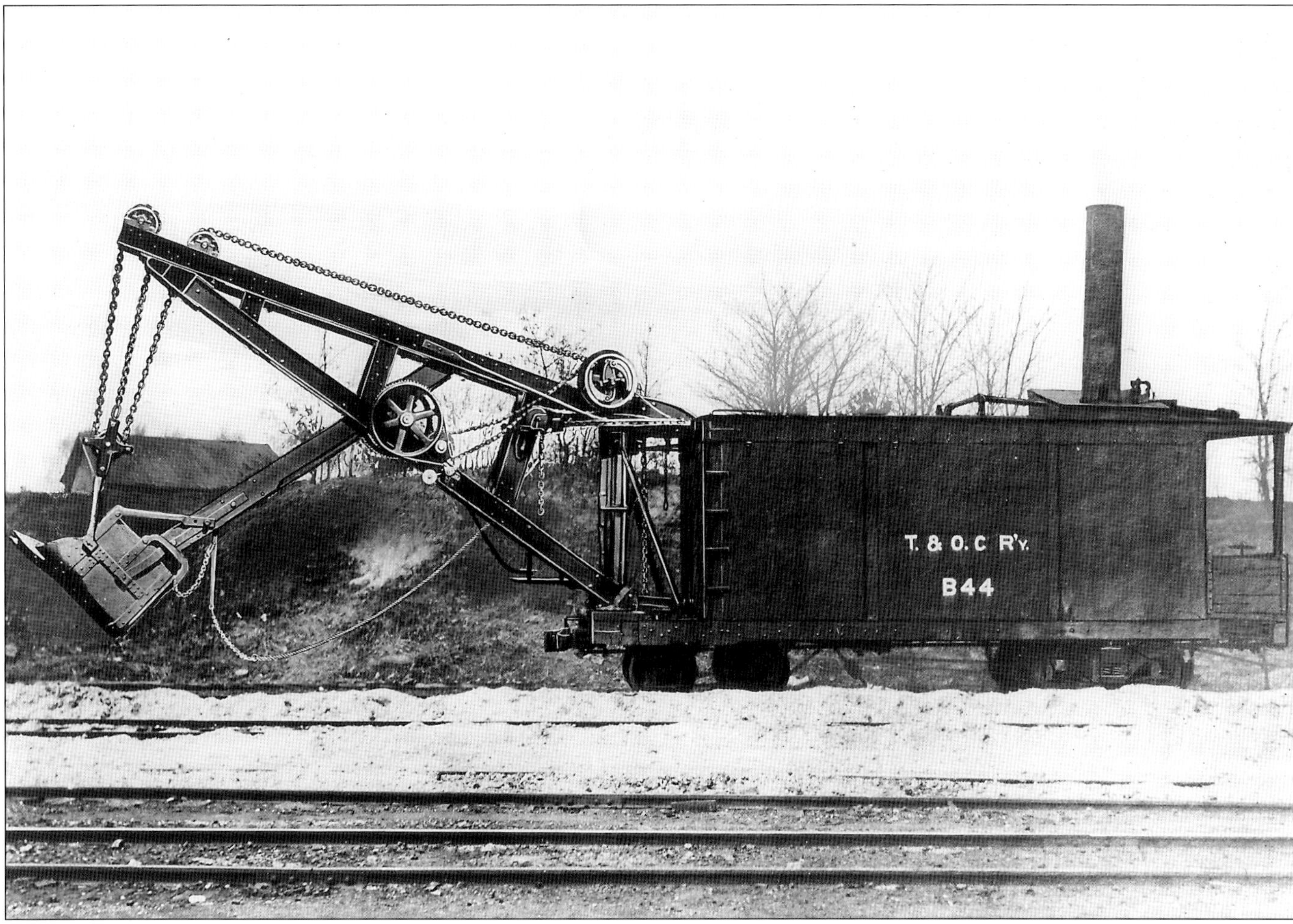

Thompson Shovel. This machine started it all. The very first excavating machine built by Bucyrus was this railroad-type machine built to the design of John Thompson, the company's Manager of Manufacturing. It was initially called the "Thompson Iron Steam Shovel and Derrick." It proved its worth for the Toledo and Central Ohio Railroad, as this picture was taken in 1910 when the shovel was 28 years old.

Bucyrus Shovel. This early "cut" of a Thompson Bucyrus shovel was called a "Combined Steam Shovel and Wrecking Car" and was described in the July 1883 edition of Railroad Gazette. It featured a vertical boiler with deck-mounted steam hoisting engines for hoisting and swinging, and a steam engine mounted on the boom for crowding.

No.0. From 1886 Bucyrus began to develop a broader range of shovel sizes. The original Thompson shovel was replaced by the No.1 at 1 3/4-yard capacity, and the new **No.2** appeared with 1-yard capacity. In 1888 the **No.3** appeared at 3/4-yard, and that same year the top-of-the-line No.0 with 2 1/4-yard capacity was introduced. Shown here the outfit weighed approximately 50 tons. Finally a small 1/3-yard shovel, the **No.4** completed the range in 1892.

No.1. Also known as the Standard 8 x 12 shovel, the 1 3/4-yard capacity No.1 was developed directly from the original Thompson shovel. The numbers referred to the dual steam engines with 8-inch bore and 12-inch stroke. This machine weighed about 40 tons in operation. By 1889, Bucyrus was fully committed to building steam shovels.

30-Ton. Beginning about 1894, Bucyrus introduced a more substantial design for its shovels. Prior shovels were designed with the swinging circle mounted overhead on top of the center post. This was known as the "crane pattern." The new shovels, with a more substantial boom, had their swinging circle at deck level. This was known as the "boom pattern." The new models were simply designated by their approximate weight, like this 30-Ton delivered to the City of Worcester, Massachusetts, in 1896.

40-Ton. The larger 40-ton railroad shovel continued the updated design introduced in the mid-1890s. This one was purchased by the Northern Pacific Railroad in 1897. Sales literature quoted, "... its design is that of employing few and massive parts instead of many connected pieces, and accomplishing the necessary movements in the most direct and simple manner." A **35-Ton** model with 1 1/2-yard dipper was introduced in 1902.

45-Ton. Not all the 77 Bucyrus shovels working on the Panama Canal were the giant 95-Ton models. This "New 45-Ton" and its larger brother, the **70-Ton**, introduced in 1903 turned out to be "Canal Specials," with 10 of the former and 35 of the latter eventually ordered by the Canal Commission. The Special 45 embodied 8-inch x 12-inch main engines and a 48-inch boiler 10 feet 8 inches long. Dipper size was 2 cubic yards.

50-Ton. This 50-Ton was shipped to the Pittsburgh Bessemer & Lake Erie Railroad in 1896. It had a similar design to the 45-Ton but carried a 2-yard dipper. While these early shovels ran on standard-gauge railroad track, the length, width and amount of reinforcement of their cars (main frames) was individually designed for each model.

60-Ton. Bucyrus introduced the 2 1/2-yard 60-Ton in 1897, then redesigned it as a 3-yard **65-Ton** unit the following year. So successful were these machines that 138 were eventually sold. This 60-Ton shovel was shipped in 1898 to the Iron Cliffs Company at Ishpeming, Michigan. Large numbers of railroad shovels found their way into the iron ranges of Michigan and Minnesota.

85-Ton. In 1899 Bucyrus announced three shovels larger than it had previously built. The 3 1/2-yard **75-Ton**, the 4-yard **85-Ton** and the flagship 5-yard **95-Ton** established the company as a serious contender in the heavy shovel market. This 85-Ton with 4-yard dipper was shipped to the Lake Superior Consolidated Iron Mines the same year it was introduced.

95-Ton. This typical scene from the Panama Canal shows one of the 32 large 95-Ton Bucyrus shovels at work on the project. The 95-Ton car measured 41 feet 7 inches long by 10 feet wide. Its main engines had a 14-inch bore and 16-inch stroke, and its locomotive-type boiler was 66 inches in diameter and 14 feet 3 inches long. The 95-Ton's standard dipper held 5 cubic yards.

30-A. This unusual 30-ton railroad shovel shows the lengths to which Bucyrus would go to design machines for special applications. All the draw works, steam engines and operator's stand were mounted on the swinging part of the machine. The front traction wheels were driven by chains while the rear wheels were steered by the hand wheel.

Sewer Shovel. This image shows an early Bucyrus railroad shovel adapted for sewer excavation. The precarious method necessitated the shovel, equipped with a special long dipper stick, being supported by the walls of the excavated trench. The dangerous operation of shoring the trench and laying the pipes underneath the machine was labor-intensive.

40-R. Beginning about 1905 Bucyrus revamped its railroad shovel line and model nomenclature for the last time. The models were henceforth designated by a suffix ‘C’ denoting chain hoist or ‘R’ for steel rope hoist. Here a model 40-R demonstrates a well-organized excavation job utilizing rail haulage in long parallel cuts. Both the 40-R and **40-C** chain type carried dippers of 1 1/2-yard capacity.

68-C. One of the most popular Bucyrus steam shovels was the 68-C with standard 2 1/2-yard dipper. In production from 1915 to 1930, and one of these was the last Bucyrus railroad-type shovel built. The 68-C boom measured 27 feet 3 inches providing a loading height of 16 feet under the open dipper door. In working order it tipped the scales at 79 tons.

68-C Crawler. When the advantages of crawler tracks on shovels became widely accepted, especially on the modern revolving shovels of the 1920s, manufacturers offered “kits” to convert the old railroad shovels from rail to crawler mountings. Here a 68-C electric shovel operates on crawlers for Sandusky Cement at Silica, Ohio.

70-C. The 70-C appeared circa 1907 with a standard dipper of 3 cubic yards and an operating weight of 84 tons. The locomotive boiler, clearly seen in this picture, measured 44 inches in diameter and 18 feet long. It steamed the main engines of 10-inch by 14-inch bore and stroke. The characters in the picture confirm this machine worked in Mexico.

78-C. Another later model Bucyrus railroad shovels was the 78-C. Introduced around 1915, it swung a standard 3 1/2-yard dipper on a 29-foot long boom to provide a loading height of 17 feet under the open dipper door. This one belonged to the extensive fleet of famous Pennsylvania earthmoving contractor Robert J. Bazley. Although the shovel enjoys complete mobility on its crawler tracks, rail transport is still in operation.

88-C. The 88-C weighed 103 tons in operation with a standard 4-yard dipper. The locomotive boiler was 50 inches in diameter and 18 feet long and, with its 12-inch by 15-inch main engines, was capable of 75,300 pounds of dipper pull. Standard boom was 30 feet which gave a loading height of 18 feet.

This railroad steam shovel, belonging to the Chicago & North Western Railway Company, demonstrates how flat rail cars were used to transport the excavated material in the era before truck transport. The cars were emptied using a "ballast unloader," an angled blade that shoved the material off the side while being pulled along the car. Vertical wooden posts placed in sockets at the sides of the cars held the unloader in place.

95-C. At 107 tons in working order, the 95-C was the largest shovel available in its day. The car-body supporting it measured 44 feet long by 10 feet wide. The 95-C was widely sold throughout the world from rock quarries in Illinois and Indiana to copper mines in Utah, and from iron ore mines in Sweden to copper mines in Spain.

103-C. One of the last Bucyrus railroad shovels to be introduced was the 103-C in 1922. It was also the first railroad shovel to be equipped with crawler tracks. The machine illustrated was owned by the Oxfordshire Ironstone Company, and uncovered iron ore in central England. It carried a 5-yard dipper on a 30-foot boom and weighed 121 tons in operation.

110-C. Billed as the largest and most powerful railroad shovel ever built when introduced about 1912, and specified for the toughest digging assignments or where blasting was restricted, the 110-C was equipped with either a 5- or 6-yard dipper and weighed 127 tons. Main engines were sized at 13-inch bore and 16-inch stroke. Conklin & Foss Company purchased this one in 1914 to tackle a tough quarrying job at Rockland Lake, New York.

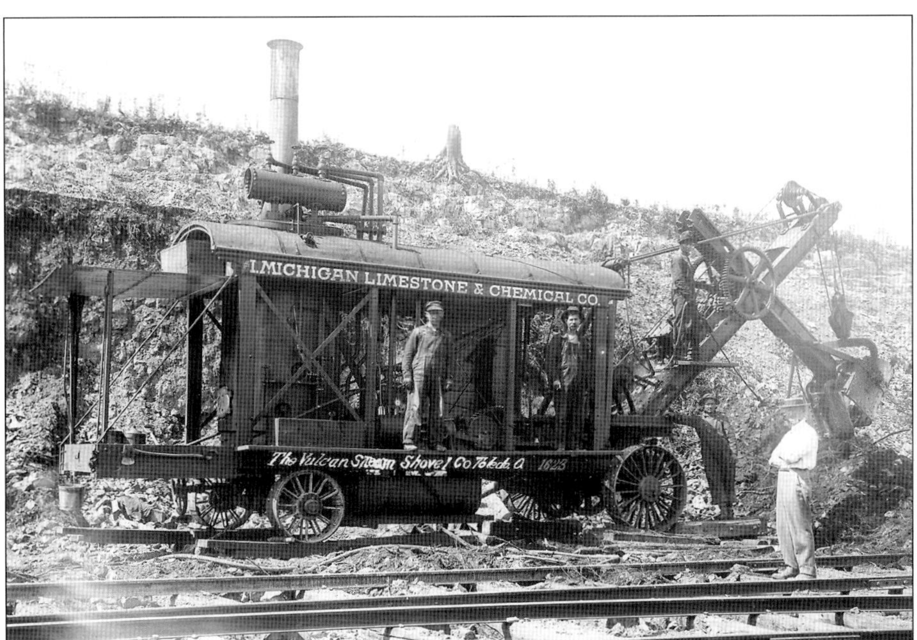

Vulcan Baby Giant. In 1911 Bucyrus acquired full patents and manufacturing rights of its long-established competitor, the Vulcan Steam Shovel Company of Toledo, Ohio. Prior to 1908 this company was known as the Vulcan Iron Works Company whose predecessor was established in 1870. In 1890 Vulcan pioneered mounting shovels on steel traction wheels like this small 1/4-yard "Baby Giant" weighing a mere 20 tons.

Vulcan Little Giant. Like the Vulcan Baby Giant, the 1 1/4-yard Little Giant in the 30-ton class was usually mounted on traction wheels. Its 7-inch-bore by 11-inch-stroke main engines provided 40 horsepower, and operating weight was quoted at 35 tons. Vulcan built and marketed an extensive line of railroad and revolving shovels which were soon integrated into the Bucyrus line after 1911.

Vulcan Giant. Vulcan's "Giant" line of steam shovels commenced in 1886 and expanded into many models including some of the largest at that time. In 1909, the Vulcan steam shovels ranged in size from 40 to 120 tons, and the forward-looking company posed some stiff competition for Bucyrus. The lattice boom design shown here is a distinctly Vulcan feature.

Vulcan Backhoe. The simplicity of the modern-day backhoe was not apparent to the early designers. Vulcan's attempt in 1896 to modify one of its "Giant" shovels with this cumbersome contraption is believed to be the first ever backhoe-type machine put in operation. Other manufacturers followed with similar designs. The simple backhoe, with its moving boom and pivoting arm, was not widely used until the 1920s.

Atlantic Shovel. In 1911 another Bucyrus competitor, the Atlantic Equipment Company was consolidated into the Bucyrus group. Atlantic was established in 1902 and had gained a fine reputation for building heavy-duty railroad-type steam shovels. They were designed by A.W. Robinson, former chief engineer at Bucyrus. The early example illustrated is working in conjunction with a car transporter in South Dakota.

Atlantic Class 25-11-1 1/4. This model was one of the smallest of the Atlantic shovels. Their model numbers indicated the three main variables in size: pounds pull, loading height and bucket capacity. Thus this shovel has a dipper pull of 25,000 pounds, a loading height of 11 feet and a dipper capacity of 1 1/4 cubic yards. Atlantic shipped 185 railroad shovels between 1904 and 1911.

Atlantic Class 45-16-2 1/2. Introduced in 1907 this mid-sized machine was one of the later models sold by Bucyrus for a short while after its takeover of the company in 1911. The 45-16-2 1/2 with 2 1/2-yard dipper weighed 73 tons in operation. Its main engines were 10-inch by 10-inch bore and stroke, and its 46-inch diameter boiler measured 20 feet in length.

Atlantic Class 80-18-4 1/2. This was the largest of the Atlantic shovels, and one that Bucyrus adopted after the company takeover. Atlantic's distinguishing and advanced features were: a very strong triangular-type boom solidly mounted on the turntable, all hoisting and crowding machinery mounted on the boom enabling a straight-line pull to the boom point, and steel ropes in lieu of chains.

Atlantic Class 80-18-3. This Class 80 has a 3-yard dipper, smaller than standard, for excavation of high-density iron ore. It is working in 1937 at Oliver Iron Mining Company's Hull Rust Pit near Hibbing, Minnesota. The unique Atlantic boom embodying all hoisting machinery can be seen in this picture. This machine has been fitted with crawler tracks which would not have been original equipment.

Chapter 7
Dredges

Dredges played a significant part in the business of Bucyrus Foundry & Manufacturing Company during its formative years. It shipped its first dredge in 1883, only one year after its first steam shovel, and continued to manufacture dredges for almost seventy years. When the company moved from Bucyrus, Ohio, to South Milwaukee, Wisconsin, in 1893, it changed its name to the Bucyrus Steam Shovel & Dredge Company to reflect the importance of both products.

By 1888 Bucyrus had shipped 14 dipper-type dredges, and their success enabled the company to design and market different types. That same year it shipped its first elevator (ladder type) dredge and, in 1890 shipped its first hydraulic cutter dredge. By 1900 the company had built dipper dredges up to 7 cubic yards, elevator dredges weighing up to 225 tons, and hydraulic dredges with twin 15-inch suction pumps. These were all monster machines in their day, and established Bucyrus as a world leader in that specialized field.

The period 1889 to 1910, when the Company shipped an average of 12 dredges per year, was the most prolific for dredge manufacture. After that time, dredge shipments leveled out to three or four per year up to World War II, but the variety and size of machines increased during that period. Most notable were two huge dipper dredges, "The Mogul" and "The Crest," built for the Great Lakes Dredge & Dock Company. The Crest, built in 1925, held the world record with its 16-cubic yard dipper, the largest mounted on any excavating machine at that time. These two floating giants served their owners well for over half a century.

Dipper Dredge "Gamboa." This was one of three 15-yard dipper dredges ordered by the Panama Canal Commission in 1913–15 to cope with massive unforeseen slides. These were the largest steam-driven dredges in the world and were arranged to operate with a 45-second cycle. They could dredge to a depth of 50 feet and ran tandem compound engines with 16-inch and 28-inch bore cylinders with 24-inch stroke.

After World War II dredge orders dwindled and only five more were sold up to 1950 when the last, a 15-inch hydraulic diesel type, was shipped to Finland. The longevity of the machines and a decrease in gold dredging contributed to market saturation. Bucyrus-Erie officially withdrew from the market in 1953 and sold its dredge business to dredge competitor Ellicott Machine Corporation of Baltimore, Maryland, in 1957. Overall Bucyrus built 344 dredges of different types.

Dipper Dredge "Crest." This was the world's most powerful dipper dredge when delivered in 1925, as noted in the introduction above. It is shown here with a 12-yard dipper for heavy digging. The Crest's diesel-electric power plant provided 1,440 horsepower for a maximum dipper pull of 350,000 pounds. Afloat she weighed about 2,500 tons. In June 1988 the Crest sank in 52 feet of water near the King's Bay Naval Base, Florida.

Dipper Dredge "Van Frank." Typical of the smaller Bucyrus dipper dredges was the 1 1/2-yard "Van Frank" sold to the United States Government in 1895. Its steam engines were 9-inch bore by 12-inch stroke, and its three spuds were hoisted by direct acting steam cylinders, held at any depth by clamps. These dredges could be purchased with either wood or steel hulls.

Leach Dredger. The first known application of the endless-track principle on an excavating machine appeared in 1905 when contracting firm Leach Brothers of Maumee, Illinois, built this prototype "walking dredge." The single endless chain of heavy planks connected by steel links was driven by a ratchet-type device operated by pulleys and powered by the dredge's main engines. A few Leach Dredges were subsequently built by Bucyrus.

Placer Elevator Dredge. Beginning in 1900 the Poutilov Works, St. Petersburg, Russia, built a number of dredges to Bucyrus designs. A ceremonial launch is taking place here. Placer dredges consisted of a bucket chain which dumped material into a hopper, a screen which received and washed the material, a stacker boom to remove the coarse tailings, and "gold-saving" tables where the gold was separated from the fine material.

Placer Elevator Dredge "Indiana No.1." Indiana Gold Dredging and Mining Company put this placer dredge to work near Oroville, California, in 1900. Its long chain of buckets each held five cubic feet. Placer dredges were normally equipped with a revolving drum screen, but in certain conditions, especially where the gold particles were very small, a shaking screen could be specified.

Placer Dredge "Yukon 3." Yukon Consolidated Gold Fields purchased this placer elevator dredge in 1905 and operated it near Dawson City. It was known as a 5-foot dredge, meaning each bucket held five cubic feet. These types of dredges were preferred by gold miners because the slow action of the buckets minimized agitation of the water, and the buckets were practically watertight, preventing loss of gold-bearing material.

Hydraulic Dredge. For the greatest output and where geological conditions allowed, the hydraulic dredge was specified. Hydraulic dredges consisted of a rotating cutter head at the end of a boom, a suction pipe of sufficient length to reach the required depth, a powerful centrifugal pump and its power unit, and a floating discharge pipe to transport the material to dump. Each dredge was specified for a particular job, and Bucyrus built them with suction pipes up to 36 inches in diameter, and discharge pipes up to four miles long.

Chapter 8
Drills

Blasthole drills are one of the three major product lines currently manufactured by Bucyrus International, Inc. These and other types of drills have sold in large volumes, and have played a major role in the company's success since the early 1930s. At that time Bucyrus-Erie, already well established in the shovel and dragline markets, realized most customers purchasing a shovel or dragline would also need a drill when blasting was required to break up the hardest material. Consequently Bucyrus-Erie acquired the manufacturing rights of the Armstrong Manufacturing Company of Waterloo, Iowa, in 1933.

As a leader in the drill manufacturing business, Armstrong had enjoyed a long and successful history. Its ancestors date back to 1868 when the Kelly, Morgan & Co. of Osage, Iowa, was formed to satisfy early drilling needs of local farmers. In 1891 that firm became Morgan, Kelly and Taneyhill Company, and then Armstrong Quam & Company in 1904. Finally in 1910 the Armstrong Manufacturing Company was established to carry on the business of these former companies.

The acquisition provided Bucyrus-Erie with a line of drills for water well and blasthole work, as well as drill tools and bit dressers. Some of Armstrong's many achievements and patents included a self-propelled drilling rig on traction wheels in 1922, a series of rubber-mounted shock absorbers in the derrick head to absorb the shock of the drilling tools, and steel-framed rigs to replace wooden frames beginning in 1927. The first Armstrong self-propelled drills with crawler tracks appeared in 1932. The drill products were sold under the "Bucyrus-Armstrong" name until 1943.

The first Bucyrus-Armstrong drills carried over from the former company were churn-types. They included water well drills (W-series), and crawler-mounted blasthole drills (T-series). These machines were forerunners of an expanded range of improved drills introduced by Bucyrus-Erie over subsequent years and incorporating the latest technology.

The principle of the churn drill, also known as a "spudder," is quite simple. A steel rod, sharpened to a chisel point, is continuously raised and then dropped in a hole, breaking up the rock, clay and sand particles. Water is poured into the drill hole to form a slurry from the churning action of the drill bit and the drill cuttings.

In 1939 Bucyrus-Erie added the first oil well drill to its roster, the L-series which remained in production until 1984. The interesting jet piercing drills, developed jointly with Linde Oxygen Company, were marketed from 1953 and found use in the hardest drilling in taconite mines.

The 50-R launched in 1952 was the first commercially successful large diameter rotary blasthole drill, and the first in a long line of R-series drills on which today's Bucyrus drills are based. The need for larger drill holes spurred the introduction of the 45-R, 60-R and 61-R drill models in the 1960s, capable of putting down holes up to 17 1/2-inches in diameter. From 1960 to 1984, the company also built rotary-type water well drills.

When the churn drill market leveled off in the 1980s, having largely been supplanted by the more productive rotary water well drill, Bucyrus-Erie sold the manufacturing rights for all its churn-type models to the Buckeye Drill Co. of Zanesville, Ohio, in 1984. It was the end of a successful product with over 11,600 sold, but it allowed the company to concentrate on design and manufacture of rotary blasthole drills for which there was a greatly expanding market in surface mines.

The R-series drill production prospered, and much innovative design work has been applied to new models introduced during the past two decades. In the mid 1980s an in-depth marketing study involving over 70 different mining companies resulted in radical design changes for drills. Introduced in 1986, the new-generation drill model 49-R incorporated a new image with a fresh outward appearance designed by an external industrial design firm.

Today the innovative 33/35/37-Series drills, at the lower end of the line, are built in India by Revathi Equipment Ltd. They combine all-hydraulic drives with the latest modular design technology.

Armstrong Drill. This early churn drill was one of the first sold by Armstrong Manufacturing Company soon after its formation in 1910. The wooden frame construction was typical of the style. The company was the successor to former drill companies going back to 1868. Bucyrus-Erie entered the drill market with the purchase of Armstrong in 1933 and acquired a line of blasthole and water well drills.

22-T. Representing the T-series crawler-mounted blasthole drills is this 22-T introduced in 1942. Comfortable drilling 6-inch holes to 100 feet, it could support a drilling string up to 1,550 pounds. A 10-horsepower electric motor or gasoline engine of 29 horsepower provided the power. The 22-T featured the patented rubber-cushioned shock absorber at the derrick head for effective use of steel cable.

27-T. When the ground surface was uneven, early drills had to be leveled and supported on planks and timbers to assure plumb and parallel holes. The 27-T was one of the first to be equipped with leveling jackscrews. The machine delivered 50 to 60 blows per minute with strokes up to 40 inches, and drill strings of up to 2,000 pounds. Diesel, gasoline or electric power could be provided. The **27-RT** was built in England by Ruston-Bucyrus.

29-T. The 26-T and 29-T were the two blasthole drills originally designed by Armstrong Manufacturing, and were adopted by Bucyrus-Erie after the takeover in 1933. The 26-T was short lived and discontinued in 1937, but the 29-T proved to be the most popular of the T-series drills with 753 sold up to 1959. It weighed 26,000 pounds and was capable of drilling holes up to 9 inches in diameter with 2,626 pounds of drill string.

42-T. The 55,000-pound 42-T was introduced in 1937 as the largest capacity portable blasthole drill on the market. It could handle 6,000 pounds of drill string and put down 12-inch holes. The illustration shows a battery of 42-Ts working for coal mining contractor Dick Construction Company near Hazleton, Pennsylvania. In 1946, the 42-T was the first to be equipped with hydraulic leveling jacks. The popular machine sold a total of 470 units until replaced by the 50-T in 1953.

50-T. The largest capacity mobile churn-type blasthole drill on the market was the 50-T. Not released until a year after the appearance of the first rotary blasthole drill (50-R in 1952), the 50-T survived until 1964, serving a niche market. With the same capacity as the 42-T it replaced, it was powered by an International 125-horsepower diesel, or 40-horsepower electric motor. It worked with a 6,000-pound drill string, lifting and dropping it 53 times per minute.

1-W. This handy truck-mounted water well drill was introduced in 1955. The fully adaptable rig could drill holes by water jetting, by use of a hollow rod drill, or as a churn drill. In addition, a rotary drill attachment was available. Depths to 350 feet could be achieved for a 2-inch hole, or to 200 feet for a 4-inch hole. The 1-W was also available as a trailer unit.

20-W. The 20-W was a portable truck or trailer-mounted churn-type water well drill that replaced the 21-W in 1949. It could handle tool weights of 900 pounds for holes up to 500 feet deep, and drill wells from 3 to 6 inches in diameter. The machine was suitable for mounting on any truck of 1 1/2 tons capacity and a wheelbase longer than 134 inches.

21-W. After the Bucyrus-Armstrong name was established, one of the first new models to appear was the 21-W water well drill. Also adaptable for exploration or blasthole work, it could be mounted on a trailer or truck greater than 1 1/2 tons capacity and 150-inch wheelbase. It was built to swing 1,000 pounds of tools for holes up to 6 inches in diameter and could drill to a maximum of 500 feet

22-W. The intermediate-sized 22-W was the most popular of the company's water well drills with approximately 3,500 sold from 1940 to 1984. It was originally designed to handle 1,500 pounds of tools and the ability to drill to 1,000 feet. The long-lived model progressed through Series II in 1964 featuring strengthened machinery allowing depths to 1,250 feet, and the Series III in 1971 with hydraulically driven derrick. It was also built by Ruston-Bucyrus as the **22-RW** with a Ruston diesel.

24-W. Produced from 1938 to 1943, the 24-W was designed on similar lines to other Bucyrus-Armstrong drills: diesel or gasoline power options for maximum mobility, trailer or truck mounting, and exclusive rubber shock absorber under the derrick crown sheave. The 24-W featured a 40-foot mast and could drill a 10-inch hole to 600 feet or a 4-inch hole to 1,500 feet.

29-W. Added to the water well drill line in 1953, the 29-W was a larger and more powerful companion to the 33-W. With a diesel or gasoline engine in the 50-horsepower range, it could handle 3,000 pounds of tools and was rated for holes to a depth of 1,500 feet. For shallow holes, diameters up to 20 inches could be drilled.

33-W. Two water well drills, the 33-W and the 38-W, were adopted by Bucyrus-Erie when it took over Armstrong in 1933. The latter could put down holes up to 4,000 feet and could work with 6,000 pounds of cable tools. The trailer-type machines included several mounted on half-track chassis. The owner of this former Armstrong No.33 boasted his truck-mounted rig drilled to 1,743 feet working with loads of over 2,200 pounds. The British version of the 33-W carried a Ford gasoline or Ruston diesel engine.

24-L. Bucyrus-Erie unveiled its first two oil well drills in 1939. The smallest was the 24-L churn-type "spudder" featuring a 40-foot standard, or 48-foot optional telescoping derrick. It was rated to drill to 2,300 feet with 1,700 pounds of tools. Its 48-horsepower engine gave it the power to hoist 1,800 pounds of tools from 1,800-foot depth in 7 minutes.

28-L. The 28-L spudder was added to the oil well drill line in 1945. With a 112-horsepower engine it was "generously proportioned" for drilling to 2,500 feet. It was offered with skid or semi-trailer mounting. Bucyrus-Erie oil well spudders featured a self-locking worm hoist for derrick raising, two-speed gear-driven main hoist, rubber cushion shock absorber atop the derrick, and an ingenious automatic spooling device utilizing a moving sheave that aligned the cable perfectly on the drum.

36-L. Released simultaneously with the 24-L in 1939, the 36-L carried a 54-foot telescopic derrick, 128-horsepower engine, and handled 5,000 pounds of tools to 2,000 feet depth. The specification also claims it could pull 4,200 pounds of tools from 3,000 feet in 8 minutes. The 36-L continued in production until 1984, long after the other models had ceased. A Series II with 60-foot derrick appeared in 1965.

60-L. Launched in 1953 the 60-L spudder was actually the last of the oil well drill models to appear. Semi-trailer, truck or skid mounted, the machine maintained most of the features of the other models except it used a manual wheel-operated system to evenly wind the cable instead of the automatic type on the smaller models. The 60-L version could drill to 1,500 feet and handle 2,750 pounds of tools.

60-RL. This was the British version built by Ruston-Bucyrus to the same capacity and specification as the American 60-L except for the power units. Standard engine for the 60-RL was a 45-horsepower Ruston diesel with later models utilizing a Dorman diesel of the same power. Standard power for the American 60-L was a heavy-duty 6-cylinder gasoline engine by Buda and others.

48-L. Contrary to what its model number would indicate, the 48-L was the largest oil well drill and the most powerful churn-type drill manufactured by the company. Released simultaneously with the 28-L in 1945, the 48-L could handle drilling to 6,000 feet, swing 6,000 pounds of tools, and lift static casing loads up to 80 tons with an auxiliary leg. Standard power unit was a 6-cylinder Waukesha gasoline engine of 219 horsepower.

26-P. Bucyrus-Armstrong churn drills were also offered specially modified for geological prospecting. Accuracy of recording data was emphasized as an important feature of these machines. Steel cable, possible only because of Bucyrus-Erie's patented rubber shock absorber in the mast sheave, improved accuracy over hemp rope formerly used. Prospecting drills were made from 1934 to 1941 and model numbers followed those of the normal churn drills except for their suffix 'P' designation.

10-R. This was a truck-mounted rotary drill adaptable for water well or shallow oil drilling, seismic work and exploration. Announced at the Tulsa Oil show in 1959 it could drill holes up to 6 inches diameter to 1,000 feet. The bit pull-down mechanism was the continuous-type, chain-driven by a hydraulic motor. The updated 10-R II launched in 1969 offered rotary drilling with air, or with a regular mud pump to depths of 1,700 feet. The **10-RH** model utilized a down-the-hole hammer drill.

30-R. Three versions of the 30-R blasthole drill were offered when introduced in 1959: truck-mounted utilizing the truck's power take-off (**30-RP**), truck mounted but independently powered (**30-RS**), and crawler mounted (**30-RC**). All drilled holes to 7 7/8-inch diameter and provided a pull-down force of 30,000 pounds through exclusive rack-and-pinion cable drive. Hydraulic cylinders leveled the machine and raised the derrick. GM or Caterpillar diesel engines of 140 horsepower were installed.

30-RS. The truck-mounted 30-RS contained its own power unit as supplied on the 30-RC model. With its 36-foot high derrick hydraulically lowered, the outfit could achieve a road speed of 40 mph. In the early 1960s the truck-mounted RP and RS models were discontinued and the 30-R II appeared in 1966. The upgraded version provided angle drilling to 20 degrees from the vertical and offered rotary or hammer drilling. Power was upped to 165 horsepower.

35-R. The original 35-R, with pull-down capability of 50,000 pounds, 6-inch to 8-inch hole diameters and machine weight of 43 tons, was manufactured in the United States and Brazil during the period 1985 to 1992. It was then upgraded to the current 35-series and joined by the **33-** and **37-Series** machines, all of which are diesel driven and currently built in India by Revathi Equipment Ltd. This series spans drilling requirements from 6-inch to 11-inch hole diameter and bit loadings from 40,000 to 70,000 pounds.

39-Series. Introduced in 1996 as the 39-R and later the 39HR, the current 39-Series drills brought revolutionary new concepts to drill design. The triangular-section mast, inspired by similar designs on Bucyrus dragline booms, was built from hollow steel tubes. All motions, including propel, rack-and-pinion hoist/pull-down and rotary head drive, are by hydraulic motors. Key specifications include pull-down capacity to 42,000 pounds, hole diameters to 8 inches, and working weight of 40 tons.

40-R. Following the success of the first large rotary blast hole drill, the 50-R, Bucyrus-Erie announced the 40-R in 1954. Designed for holes from 6 3/4 to 9 inches, it could be furnished with either a Caterpillar 215-horsepower diesel engine or straight electric motor of 150 horsepower. Both the 40-R and 50-R utilized a rack-and-pinion hoist/pull-down system, a technology borrowed from the crowd systems on the company's shovels. The 40-R tipped the scales at 50 tons.

45-R. The 45-R became the standard blast hole drill for medium-sized mines in the 1970s and 1980s. First appearing in 1966, the 75-ton 45-R drilled holes to 10 5/8-inch diameter with bit loading to 70,000 pounds. Electrically powered but with a diesel option, it featured a 50-horsepower DC electric rotary drive, twin hydraulic cylinders to raise and lower the mast, hydraulic leveling jacks, and an exclusive hydraulic pipe handling system that changed drill pipes without the operator leaving his cab.

47-R. The 47-R embodied the modern mechanical and electrical attributes of the 45-R, but the heavier 125-ton machine achieved the same capacity and a faster drilling rate than the 50-R.which it replaced. Holes up to 12 1/4 inch could be drilled, and pull-down loading was upped to 90,000 pounds. Its Allis-Chalmers on-board air compressor put out 1,450 cubic feet per minute. The 47-R was built from 1983 to 1991.

49-Series. Introduced as the 49-R in 1986 with bit loading to 100,000 pounds, the first in this series featured a new chainless pull-down system with DC motor mounted on the head and hydrostatic propel drive with planetary gear reduction in each crawler frame. Other 49-Series models included the **49-RH** (1990) and **49-R II** (1992) with bit loading increased to 120,000 pounds, hole size to 15 inches and machine weight to 165 tons. The 49-R III and current 49RH drill holes to 16 inches with 141,000 pound bit loading, and state-of-the-art programmable logic controller.

50-R. The 50-R introduced by Bucyrus-Erie in 1952 became the first commercially acceptable large diameter rotary blasthole drill. It utilized twin air compressors, each powered by its own 125-horsepower electric motor, to remove the cuttings by air instead of mud. Main hoist and rotary drives were 50-horsepower electric motors with Ward-Leonard control. The 56-ton unit was mounted on heavy-duty 42-inch wide crawlers and could drill holes up to 12 1/4 inches in diameter.

55-R. Making its entry in 1972, the 55-R drill earned a strong reputation in open-pit copper and coal mines as well as limestone quarries. With electric or diesel versions weighing just over 100 tons, the 55-R featured a box-type mast and AC motor drives for all motions. Bit loading to 100,000 pounds and holes to 12 1/4 inch could be handled.

59-Series. The 59-R followed in the footsteps of the well proven 49-R when it appeared in 1992. Currently Bucyrus' flagship drill, the massive 59-R weighs up to 200 tons and produces 17 1/2-inch holes with a bit loading of 165,000 pounds. Electric DC drives are employed for rotation and main hoist, with AC drives for air compressor and hydraulics. The 59-Series drills have found their main use in taconite mines where the hardest drilling is encountered.

60-R III. The 60-R and its dual motor rotary drive and dual air-compressor brother, the 61-R, became workhorses of the surface mining industry and the company's biggest selling large blast-hole drill. Between 1962 and 1993, 374 of the 60-R/61-R drills were produced. The first 60-R could drill holes to 12 1/2 inches in diameter with bit loading to 90,000 pounds. Diesel or electric power was available. Upgrades brought the Series II in 1968, Series III in 1973 and Series IV in 1979 with bit loading to 125,000 pounds.

61-R. The first 61-R was delivered to Peabody Coal's River Queen Mine in Kentucky in 1962. Physically similar in size to the 60-R, its increased capability came from doubling up air compressors at 200-horsepower each, and doubling DC rotary motors at 70 horsepower each. Holes to 17 1/2 inches could be accommodated and bit loading was enhanced to 130,000 pounds.

67-R. Two massive 67-R drills were shipped to the Moura Coal Mine, Queensland, Australia, in 1986. They were the largest drills made by Bucyrus at that time, but only these two were produced. Based on the time-proven 60-R, the 67-R was the "super drill" of the 1980s. Its high productivity, with bit loading to 160,000 pounds and 17 1/2-inch hole capability, was matched with operator comforts including hot plate, refrigerator, microwave, and washbasin.

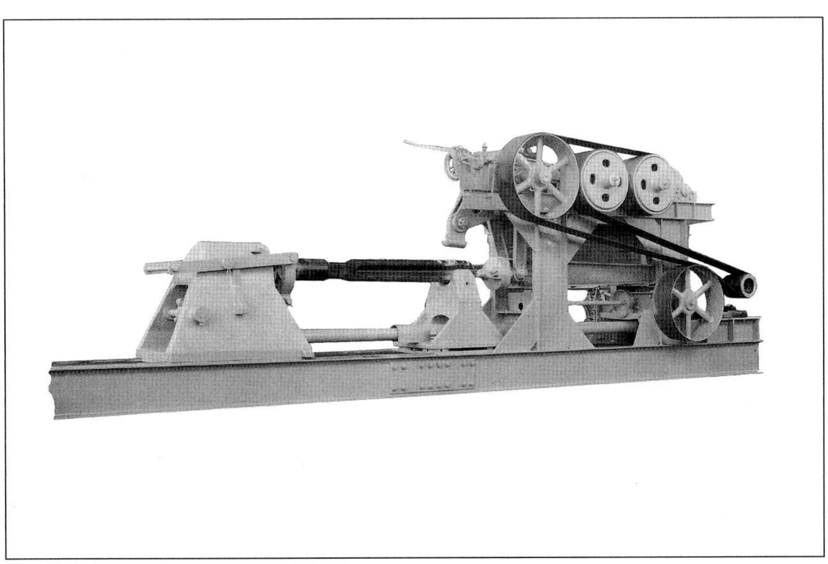

No. 12 Bit Dresser. Churn drill bits required continual resharpening via a forging process. For years bits were heated on site by use of a portable furnace and then hand dressed with a sledgehammer. The bit dressing machines inherited from Armstrong in 1933, and continued by Bucyrus-Erie until 1967, provided an efficient means of sharpening bits far more accurately than could be accomplished by hand. Bit dressers came in two sizes, No. 8 for bits 4 1/2 to 6 inches, and No. 12 for bits 6 to 12 inches in diameter.

JPM-3. Drilling in extremely hard taconite ores presented special challenges to the drills available in the 1950s. Consequently Bucyrus-Erie and Linde Oxygen Company jointly produced a jet piercer drill that utilized a fuel oil and oxygen flame at the end of the drill pipe that heated the rock. A water spray applied at the bottom of the drill hole caused the heated rock to spall and be blown out of the hole. Although drilling rates were drastically improved, fuel costs were extremely high.

Winkie Drill. The smallest Bucyrus-Erie drill product was the Winkie drill, purchased from its inventor, Al Wink. The hand-held diamond-core drill was ideal for marine blast hole drilling, obtaining core samples from reinforced concrete, and coring holes through concrete walls for gas, water and electrical services. From 1959 to 1961, 163 were sold.

120-SL. The short-lived 120-SL was a mobile, self-contained well-servicing machine with highway mobility to 40 mph. Its 67-foot mast had a working capacity of 180,000 pounds, and its massive winding drum held enough rope to perform rotary workovers for holes to 12,000 feet deep. Line speeds ranged from 86 to 1,000 fpm. Although displayed at the Tulsa Oil Show in 1959, the 120-SL was produced for only a short time after that.

2400-R. Bucyrus participated in the rotary water-well drill market and produced 260 R-Series type rotary water-well drills: models 10-R, 2400-R and **2450-R** from 1960 to 1984. These models were highly mobile, truck-mounted outfits with modern hydraulic control and a host of engine and compressor combinations. The 2400-R, which appeared in 1976, could put down a 6-inch hole to 2,500 feet, or a shallow hole up to 30 inches in diameter.

Chapter 9
Tractor Equipment

In 1935 Bucyrus-Erie announced its entry into making tractor equipment, and the following year became the major supplier of tractor equipment to International Harvester Company. This was a win-win arrangement for both companies who were just pulling out of the Great Depression of the early 1930s. Bucyrus-Erie's big plants at South Milwaukee, Erie and Evansville were operating at about one-third capacity and welcomed the extra work. International was planning to expand its manufacturing capacity and was about to launch a new range of crawler tractors, but didn't want to utilize valuable space for major welded components. It recognized Bucyrus-Erie's quality manufacturing capability, and International's dealers would be pleased to have a reliable source of supply.

The two companies' engineers worked closely together to design attachments for new tractor models before they were launched. Equipment produced comprised 2-wheel and 4-wheel pull-type scrapers, hydraulic- and cable-operated bulldozer blades, cable winches, rippers, and loaders. Bucyrus-Erie adopted the names "bullgrader" to describe a blade whose angle could be adjusted (angle dozer), and "bulldozer" to describe a fixed-angle or straight blade. Bucyrus-Erie also produced bulldozer blades and winches for Allis-Chalmers, Caterpillar and Cletrac.

During Bucyrus-Erie's involvement with tractor equipment, three distinct designs of 4-wheel scrapers were produced. The first machines were controlled by a clever single-cable arrangement that operated all three motions, bowl raise, apron and ejector, but it was not as successful as expected. The more-successful S-series scrapers followed. These were operated from a two-drum winch with the bowl raised from the rear instead of the more common front-lift arrangement. Finally in 1948, the first of the B-series appeared, featuring the more-conventional front-lifting bowl.

During the World War II years, 1942 to 1945, some two thirds of the Company's production was devoted to the war effort. A large percentage of this production was equipment for the thousands of tractors delivered to the armed forces around the world. The bulldozers proved indispensable in the construction of access roads and airstrips from the Alaskan Aleutian Islands to the South Pacific, as well as in the allied countries of Europe.

Bucyrus-Erie continued producing tractor equipment after the war, and its new B-series earned a good reputation. But in 1953 International Harvester announced it intended to produce its own allied equipment commencing the following year. Consequently, after some 20 profitable years as a tractor allied equipment builder, Bucyrus-Erie agreed to phase out these products.

Bucyrus-Erie's first scrapers were four-wheel types operated by a single cable from a winch on the tractor. An ingenious linkage and sheave arrangement worked all three movements: bowl hoist, apron and ejector. They were built in 4 1/2-yard and 7-yard sizes and worked well behind International's crawler tractors. But the design proved too complicated from maintenance and reliability standpoints, and consequently was soon replaced by double-rope models.

S-67. The successful S-series scrapers replaced the early single-rope models. They were operated from a two-drum winch, but the bowl was lifted from the rear instead of the more usual front lift arrangement. The initial range comprised the S-45, S-67, S-90, S-112, and S-152 with heaped capacities listed as 5, 7 1/2, 10, 12 1/2, and 17 cubic yards respectively. The S-67 was designed for tractors ranging from 45 to 55 horsepower.

S-112. The S-112 matched tractors in the 70- to 95-horsepower range. It weighed about 8 tons empty, and 167 feet of 1/2-inch cable was required to reeve the sheaves for its three motions. In 1948, Bucyrus announced upgrades for the S-series scrapers excluding the S-152. Improvements included a curved "goose neck," improved straight-line cable reeving, adjustable-height push bumper and double-curved cutting edge. Capacities remained the same but models were renumbered S-46, S-68, S-91 and S-113.

B-250. In 1948, Bucyrus-Erie released the first of its B-series pull-type scrapers, the 27 1/2-yard heaped capacity B-250, designed to match the new record-beating International TD-24 tractor. It was the first of Bucyrus-Erie's more conventional front-lifting bowl scrapers. Its scale weight was approximately 20 tons, and 283 feet of 9/16-inch cable was required when fully rigged.

B-170A. The B-170, rated at 19 yards heaped capacity for tractors of 75 horsepower and up, followed hot on the heels of the B-250. By 1952 the B-series was completed by the addition of the 10-yard B-91 and the 14-yard B-113 which replaced the former S-series. In 1952, the B170 was upgraded to the larger B-170A with 10% increase in weight, larger tires, and 21-cubic-yards heaped capacity.

B-113. Like other Bucyrus-Erie scrapers, the B-113 was of open bowl design. Cable reeving was simplified and when cutting material the curved apron and bowl design produced a "boiling" action breaking up chunks and filling air voids. The curved ejector was hinged at the front of the bowl and rolled the material out when lifted. Bucyrus-Erie ended its association with tractor equipment in 1954 when International acquired the manufacturing rights and started building them itself.

G-38. Bucyrus-Erie's 2-wheel scrapers originated in 1937 and two sizes were marketed at that time: the H-44 and H-74 of 5- and 8-cubic-yards capacity. They were rear dumping and hydraulically operated through hose connections to a hydraulic pump and valve system on the tractor. Upgrades brought replacement models after World War II. The G-38 and G-58 carried 4 1/2 and 6 heaped cubic yards respectively. Two-wheel scrapers were discontinued by 1951.

TD-18A Bullgrader. Designed specifically to match International Harvester tractors, the Bucyrus-Erie dozer blades were available with hydraulic or cable control and in sizes to match each tractor model. This TD-18A is equipped with a 12-foot-wide hydraulic "Bullgrader," Bucyrus-Erie's term for an angle dozer. Mounted on a C-frame, it could be angled left or right. Hydraulic pressure was a low 200 psi.

TD-18A Bulldozer. The more-simplified straight "Bulldozer" blade is mounted here on an International TD-18A tractor. The straight blade could perform a smaller variety of tasks but its simple and robust design made it more suitable for heavy bulldozing. The TD-18A Bulldozer is 10 feet wide, some two feet narrower than the Bullgrader version.

Wartime Equipment. This field of International crawler tractors, most fitted with Bucyrus-Erie equipment, gives some idea of the thousands of tractors sent overseas to serve the armed forces in World War II. Although the company built gun carriages for the war, it was fortunate in that the bulk of its wartime contracts were filled with its regular products, such as tractor equipment and cable excavators, avoiding costly retooling.

TD-24 Cable Bulldozer. The largest tractor to be equipped with Bucyrus-Erie equipment was International's TD-24. First announced in 1947, it was for a time the world's largest crawler at 180 horsepower. The straight cable bulldozer shown measures 12 feet wide and is an optional "tilting type." The left-hand blade brace could be manually adjusted to raise one side of the blade up to 13 inches above the other side.

TD-6 Dozer Shovel. The Bucyrus-Erie hydraulic "Dozer Shovel" was made for the International TD-6, T-6, TD-9 and T-9 tractors. The first version from the early 1940s, shown here on a TD-6, incorporates vertical lift cylinders with the arms guided by a vertical track to gain longer reach. No hydraulic power was provided to the bucket—it dumped by gravity when a latch was released. The TD-6 bucket held 1/2 cubic yard.

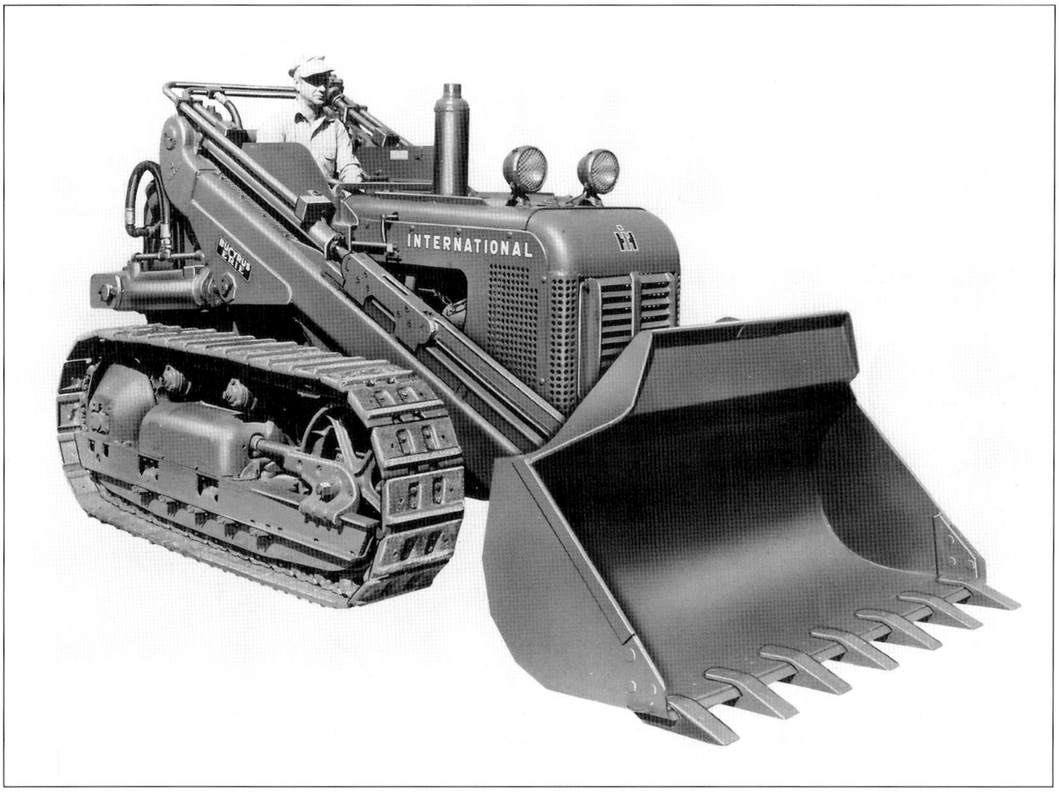

TD-9 Dozer Shovel. The later "Dozer Shovel" design, appearing in 1949, is recognized by its straight arms and under-slung hoist cylinders. It sported hydraulic bucket control, absent on the earlier models. Bucyrus-Erie's "PVT" (Pump, Valve, Tank) hydraulic system operated between 400 and 650 psi. Standard bucket on the TD-9 tractor illustrated was 1 cubic yard.

Rollers and Rippers. Bucyrus-Erie's tractor equipment included various types of winches, tractor-mounted push plates, rooters, rippers and rollers. Tamping rollers, with different feet arrangements, were 9 ft. 4 in. wide, and weighed up to 9,250 pounds when ballasted with sand. Heavy-duty cable-operated, towed rippers featured a simple trunnion hoist arrangement that provided maximum ripper shank penetration. Model **CR-3** matched tractors up to 100 horsepower; model **CR-4** served the largest tractors.

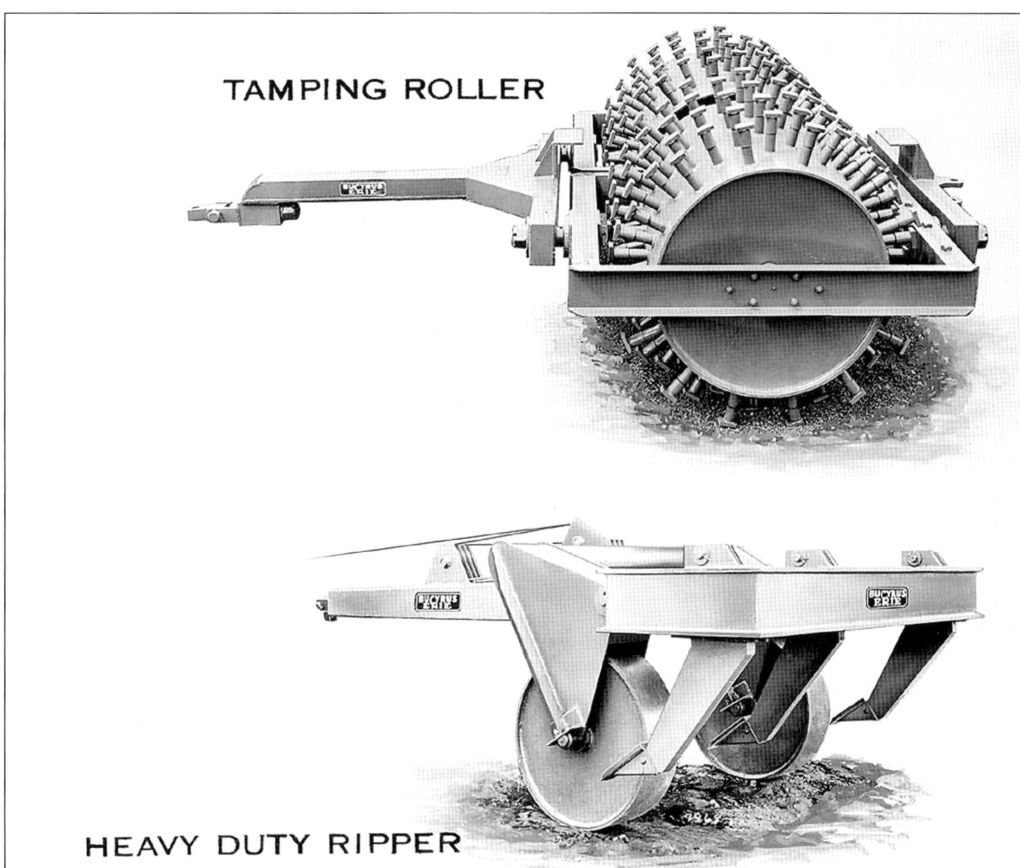

Chapter 10
Hydraulic Excavators

Bucyrus-Erie made its first move into hydraulic machines by purchasing the Milwaukee Hydraulics Corporation in 1948. Two years earlier this company had introduced one of the world's first hydraulic cranes called a Hydrocrane. The truck-mounted hydraulic crane was soon developed into a hydraulic backhoe and given the logical name Hydrohoe. This early line laid the foundations and provided experience for the company to launch a new line of crawler-mounted fully revolving excavators in the 1960s. First came the 3-RB, designed and built in England by Ruston-Bucyrus Ltd., then in 1965 Bucyrus-Erie unveiled the 20-H, first of a new range of excavators that would eventually cover sizes from the 12-ton 15-H, to the 60-ton 40-H that could handle buckets up to 4 cubic yards.

1975 saw the introduction of the 350-H, heralding a complete new range of hydraulic excavators. These state-of-the-art machines, with high-pressure piston pumps replacing the outmoded low-pressure gear pumps in the former models, garnered Bucyrus-Erie an enviable reputation in the hydraulic excavator field. Meanwhile the truck-mounted Hydrohoe and Hydrocrane products were upgraded over the years and the H-3 and H-5 Series II models survived until 1981. Unfortunately the hydraulic excavator line fell victim to the drastic recession of the mid-1980s, and all products were discontinued by 1985.

H-2 Hydrocrane. The original truck-mounted 2-ton Hydrocrane is shown here working with a hydraulic clamshell. The single hoist line is raised and lowered by a pair of hydraulic cylinders working between two sheave blocks with multiple parts of rope. The truck engine also drives the hydraulic pumps by means of V-belts from a power-take-off on the truck transmission. With standard International L-152 truck, the excavator rating was 1/4 cubic yard, and the entire outfit weighed 11,100 pounds.

H-3 Hydrohoe. In 1951 Bucyrus-Erie introduced the Hydrohoe and offered it on the H-3 model which, like the H-2, was previously only available as a crane. It worked the same way as the Hydrocrane, but a hydraulic cylinder was mounted on the upper part of the boom to work the pantograph linkage as shown in the photograph. Rated at 3/8 cubic yard, the H-3 graduated to a Series II in 1960, with extended ranges. Weight including truck was approximately 17,600 pounds.

H-4 Hydrohoe. One of the models adopted from the Milwaukee Hydraulics Corporation was the H-4. But although this 4-ton model boasted an independent engine to power the hydraulic pumps, it was not continued. Instead the independent motor feature was designed into H-3 models by 1956 and in all subsequent Hydrohoe models. An H-4 Series II, as a crane only with lifting capacity to 10 tons, was offered for a short time in 1967.

H-5 Hydrohoe. The 1/2-yard capacity H-5 appeared in 1955 with a choice of Ford gasoline or Hercules diesel engines in the 100-horsepower class. With a more conventional hydraulic backhoe configuration, it retained the vertical cylinders and sheave arrangement for conversion to crane. Swing motion on all Hydrohoe models was 360 degrees, but not continuous, and achieved by two hydraulic cylinders and pulley system. The 15-ton capacity **H-6** model was offered as a crane only for a short time in 1967.

3-RB. With the hydraulic excavator phenomenon gaining momentum faster in Europe than in America, it was no surprise that Ruston-Bucyrus designed and built the company's first fully revolving hydraulic crawler machine at Lincoln, England. The 1/2-yard 3-RB was launched in 1962 with advanced features such as independent hydrostatic motors for each crawler, continuous 360-degree swing with ball-bearing swing circle and hydraulic motor, and variable-stroke piston pumps operating at a high 2,400 psi.

3-RB II. While the first model 3-RB sold some 70 units, it was not considered a resounding success. With minor upgrades, new cab, and weight increased by about one ton to 11 tons, the 3-RB Series II was announced in 1965. A wheeled version was also available. Digging depth with standard arm was 14 feet, and its Ruston 3YDAN diesel engine put out 55 horsepower. The 3-RB was discontinued when the American-designed 20-RBH commenced British manufacture in 1967.

15-H. Bucyrus-Erie unveiled its smallest hydraulic excavator, the 1/2-yard 15-H, in 1966. It featured constant-flow gear-type hydraulic pumps to operate the boom and bucket, and a hydraulic motor providing continuous 360-degree swing. Hydraulic motors mounted in each crawler assembly provided independent steering with direct gear reduction to the crawler tumblers. Optional engines in the 74-horsepower class were available, and the machine weighed 24,400 pounds in standard configuration.

15-H Truck. A few 15-H excavators were sold as truck-mounted units with the purchaser usually providing his own truck. Although digging depth was slightly reduced, mobility was the advantage. The 15-H was upgraded to the 15-H Series II in 1969 with similar ranges and engine options, and weight upped to 25,300 pounds. It remained in production until 1976.

15-RBH. This is the British version built by Ruston-Bucyrus from 1969 to 1972. Known as the 15-RBH, its design features and specifications were very similar to its American counterpart, except it ran on a British-built Perkins 6.354 engine of 94 horsepower. Tractor-type crawlers with grousers became standard by 1972, which increased operating weight to 27,750 pounds.

20-H. Bucyrus-Erie's all-new hydraulic excavator line was launched in 1965 with the 20-H, shown here with a special rotary boom. The 20-ton 3/4-yard capacity machine employed gear-type hydraulic pumps with Cummins or GM engines of 106 horsepower. Initially the undercarriage was similar to a cable-operated excavator, with drive through the machine's center shaft from a deck-mounted hydraulic motor and steering by air-operated clutches and brakes. Upgrades brought changes through Series II to Series IV. The 20-H Series III, introduced in 1972, boasted independent propel motors mounted in the crawler frames.

20-RBH. The British 20-RBH commenced production in 1967 with almost identical design and specification to the 20-H. It retained the crawler undercarriage (based on the Bucyrus-Erie 25-B excavator) with air-actuated clutches and brakes for steering. Toward the end of its production life in 1972, the 20-RBH was available with tractor-type crawlers, adding approximately 2 1/2 tons in weight. In North America a tractor-type "Hi-Walker" version was marketed as the 21-H.

30-H. The 30-ton 30-H appeared in 1967 with standard dipper of 1 1/4-yards capacity and choice of Cummins or GM engines of 170 horsepower. Digging depth was 23 ft. 6 in. with standard boom or 28 ft. 9 in. with long arm. The machine illustrated was built in Bucyrus-Erie's Canadian plant at Guelph, Ontario, which manufactured machines from 1957 to 1971.

30-H II. The 30-H could be supplied with the tractor-type "Hi-Walker" as shown. The Series II version came out in 1969 with increased weight and more powerful engine options to 230 horsepower. The 30-H remained in production until 1979 but, unlike its smaller 20-H stablemate, it never achieved a modern undercarriage with independent propel motors in the crawler assemblies.

30-RBH. The 30-RBH was built at the Lincoln, England, factory from 1969 to 1974. It had the same standard GM 6V-53 diesel as in the American machine but rated at 178 horsepower. It also embodied a similar propel drive, with two hydraulic motors mounted on the revolving frame driving the propel gearing through the machine's center shaft. Ranges and digging depths varied slightly from the American machine.

31-H. In 1973 Bucyrus-Erie introduced a special pipelining hoe called the 31-H, designed to capture some of the Alaska pipeline work going on at that time. Based on the 30-H, it moved on the well-proven "Hi-Walker" undercarriage, but was driven from the upper works and steered using air-operated clutches and brakes. Its standard 1 3/4 dipper could reach down over 19 feet, and out to 31 feet at ground level.

40-H. With standard dipper of 3 cubic yards, the 40-H topped out the first series of hydraulic excavators in 1970. Unlike its model number would suggest, the 40-H weighed some 58 tons in operation. Hydraulic gear pumps driven by a 280-horsepower Cummins NT-855 engine provided the power, and independent propel motors were embodied in each crawler side frame. The machine illustrated is equipped with a special grapple for handling pipes or blocks.

40-H IPA. IPA stands for "Increased Production Attachment." Introduced in 1974, the 61-ton machine was Bucyrus-Erie's answer to clean coal loading. The short-coupled boom and handle allowed dippers up to 8-cubic-yards capacity with digging depth to 21 feet—some 6 feet shorter than the standard machine. Ball-bearing swing circles were a feature of all Bucyrus-Erie hydraulic excavators.

65-HL Logger. Based on the 40-H, the 65-HL was a special heavy-duty version with log grapple replacing the bucket. The ball bearing swing circle and heavily guarded lower works propel stood up well to tough conditions in the woods. The operator's cab was raised above house roof elevation for 360-degree visibility, and safety glass used throughout.

150-RH. About the same time Bucyrus-Erie introduced the first of a complete new line of hydraulic excavators in 1975, Ruston-Bucyrus launched one of its own. Designed and built with European influence at the Lincoln factory, it would be sold in Europe as well as the United States. Variable-flow piston pumps operating at a high 3,500 psi were at the heart of its hydraulic system. Complementing the B-E line at the lower end, the 16-ton 150-RH backhoe could dig to 16 ft. 9 in.

150-RHW. As well as this wheeled version capable of a top speed of 13 mph, the 150-series excavators were offered with a variety of boom and arm configurations and a host of attachments. Standard bucket was rated at 3/4 cubic yard. A choice of engine options of 70, 85 or 97 horsepower was provided by Ford, Deutz or Perkins. The 150-RHW weighed approximately 17 tons and could reach over 33 feet with extended ditching attachment.

155-RH. The 155-RH replaced the 150-RH in 1980. It was upgraded from the former model with minor specification changes including increased hydraulic pressure to 3,750 psi. These machines found favor with drainage and river authorities in Europe. A choice of "mono-bloc" single-piece or two-piece pinned booms gave plenty of opportunities for special jobs. A clamshell attachment specified on a long boom extension could dig down over 40 feet.

175-RH. Another Ruston-Bucyrus special was the 175-RH. It was actually a hybrid machine consisting of the upper works from the 155-RH and the lower works from the 220-RH. This resulted in a low-ground-pressure machine with similar digging ranges to the 155-RH. Only three units of this model were sold.

220-RS. The 220-RH backhoe was launched soon after the 150-RH in 1976 with the same design features as the smaller model. Early machines came with a Ford 2715E diesel engine, but later versions carried the Deutz air-cooled F6L912, developing 106 horsepower. The shovel version, also known as the 220-RS, offered a choice of 54-inch or 69-inch wide coal-loading dippers of 1 1/2 and 2 cubic yards respectively. The 220-RS weighed 23 tons.

300-H. The 300-H and 325-H were added to Bucyrus-Erie's all-new line of excavators in 1977. The 1-yard 300-H shown is equipped with an amphibious undercarriage manufactured by Kori Corporation of Harvey, Louisiana. The pontoons were made of "Cor-Ten" steel having 5 to 8 times the atmospheric corrosion resistance of structural carbon steel. Drive cleats were made of extruded lightweight alloy aluminum, and the entire outfit on land exerted a ground pressure of only 1.4 psi.

325-H. This intermediate-sized machine was part of Bucyrus-Erie's advanced line of excavators announced in 1975 after many years of research and development. It featured piston-type propel motors in each crawler assembly and a stronger, stress-flow designed carbody. The 35-ton 325-H worked with a 1 3/4-yard standard hoe dipper and was powered by GM or Cummins diesel engines in the 210-horsepower class.

350-H. In 1975 Bucyrus-Erie introduced a brand new breed of hydraulic excavators with the 350-H. These machines were state-of-the-art, with high-pressure variable-displacement piston pumps operating at 5,300 psi and "demand/control" hydraulics that automatically proportioned speed with optimum force during each phase of the digging cycle. The sleek machinery house was made of "Tri-Armor" reinforced fiberglass with integrated color. The 350-H was in the 50-ton class and carried a 2 1/2-yard standard hoe dipper.

375-RS. The 375-RS was built in Lincoln, England, and sold as a Ruston-Bucyrus machine from 1981 to 1983. Derived from the 350-H, it possessed identical mechanical and hydraulic features, except for a cross-roll swing circle instead of the ball-bearing type. Available only as a face shovel, it was offered with front-dumping or clam-type dippers of up to 4 cubic yards. Power came from a GM 8V-71N diesel of 275 horsepower, and the machine weighed approximately 55 tons.

400-H. In 1979 Bucyrus-Erie extended its hydraulic excavator range upwards with the introduction of the 400-H and 500-H backhoes. They retained all the advanced features of the smaller models except their crawler drives were the non-tractor type, with a hydraulic motor in each crawler frame driving the tumblers through a planetary reduction.

The 400-H, in the 80-ton weight class, came with a standard 4 1/4 dipper. A Detroit 8V-92T put out 392 horsepower.

500-H. The big 500-H tipped the scales at 115 tons and was powered by a 603-horsepower Detroit 12V-92T engine. It could reach down over 40 feet with an optional long arm, and over 60 feet horizontally at ground level. Like the earlier 40-H machine, the 500-H could also be equipped with the IPA (Increased Production Attachment). Digging depth was reduced to 28 feet, but coal dippers rated up to 18 cubic yards could be handled.

550-HS. This was not merely the shovel version of the 500-H; the entire machine was beefed up for heavy-duty mining service. The longer and wider undercarriage incorporated cast steel shoes with 1 1/2-inch grousers, and the swing circle was beefed up to a double-roller type. The 140-ton 550-HS with 10-yard clam-type dipper was Bucyrus-Erie's largest hydraulic excavator. In production for only two years, it suffered the same fate as all other products from the Construction Equipment Division when they were discontinued by 1985.

60-S. This unique hydraulic swing shovel was marketed only in 1970. It was designed to combine the advantages of a wheeled loader and a hydraulic excavator. Its high-speed and flexible undercarriage (resembling an army tank) gave it superior mobility, while its 360-degree swing and high crowding capability mirrored that of a shovel. The installed 380-horsepower Cummins NT-380 engine gave it a top speed of 6.7 mph. The machine weighed 101,000 pounds and the bucket was rated at 5 cubic yards.

Chapter 11
The Dynahoe

In 1971 Bucyrus-Erie Company acquired the Hy-Dynamic Company of Lake Bluff, Illinois. This company was best known for its "Dynahoe," a tractor-mounted combination loading shovel and hydraulic backhoe (loader-backhoe), but it also provided Bucyrus-Erie with a line of rough-terrain, wheel-mounted hydraulic cranes (see Chapter 12) all of which became Bucyrus-Erie products. This acquisition came at a time when Bucyrus-Erie's Construction Equipment Division was at the height of its activity and demonstrated the company's desire to capture all segments of that market.

Competitive loader-backhoes originated as agricultural wheel tractors and were generally much smaller in capacity and lightly built. But the unique Dynahoe was heavily built from the ground up as an excavating machine, and the largest of its type on the market. The average loader-backhoe in the early 1970s had a digging depth from 14 to 17 feet, 80-horsepower engine, and weighed 10 tons. However the largest Dynahoe boasted a digging depth of 20 feet, carried a 155-horsepower engine, and weighed 16 tons. An independent survey of construction equipment owners conducted by *Construction Methods* magazine showed the Dynahoe as the number one preference in the United States.

Although not selling in such huge quantities as the smaller competitive machines, the Dynahoe served a niche market and garnered sufficient customers to keep the machines in Bucyrus-Erie's product line until the company decided to dismantle and sell its entire Construction Equipment Division in 1985. Manufacturing rights and designs of the Dynahoe were sold to former company employees who formed a new company, Bucyrus Construction Products (BCP) based at Erie, Pennsylvania. The following year BCP became a division of Terex Corporation.

Model A. The Hy-Dynamic Company introduced the Dynahoe loader-backhoe in 1959. One of the first models was this Model A equipped with a Continental F-244 gasoline engine developing 65 horsepower. The Model AD was also offered with 65-horsepower Continental HD-277 diesel engine. Maximum digging depth was 13 feet and backhoe loading height 12 feet. Operating weight was 13,000 pounds. Even this early design sported a neat efficient backhoe linkage, integrated design and massive frame.

Dynahoe 140. Dynahoe upgraded its models in 1966 and launched a new range of machines with a 3-digit nomenclature, the first two digits indicating digging depth in feet. Models A and AD were replaced by the model 140, and other models in the new lineup included the short-lived 120 and 130 (discontinued by 1968), and the larger 160. These models, plus the 190 and 200-4 introduced in 1968, were current when Bucyrus-Erie took over the company in 1971.

Dynahoe 140B. The Dynahoe name change to Bucyrus-Erie in 1971 also heralded updated B-series machines. All were equipped with GM Detroit diesel engines and featured advanced hydraulics and increased operating weights. The 140B was powered by a 90-horsepower, 3-cylinder Detroit 3-53 diesel engine with torque converter. The backhoe could reach to 22 feet at ground level and the loader bucket held 1 1/4 cubic yards. Machine weight was 14,800 pounds.

Dynahoe 190B. The larger 190B came with the same Detroit 3-53 diesel as in the 160-B, but rated higher at 95 horsepower. It was basically the same machine as the 160B but with greater backhoe digging depth to 19 feet and loading height increased to 12 feet 9 inches. Front loader capacity for both machines was 1 1/2 cubic yards.

Dynahoe 190-4. This was the 4-wheel-drive version of the Model 190, easily recognized by its four equal-sized tires. A 4-cylinder Detroit 4-53 diesel of 126 horsepower provided the power. Steering was provided to the rear wheels only, while the front axle was free to oscillate. The heavier machine, at 22,750 pounds including ROPS cab, could support a loader bucket of 1 3/4 cubic yards.

Dynahoe 190-4 Side Dump. Typical of the many optional attachments for both the front and rear of these machines was the side-dumping bucket. Shown is the 1 3/4-yard bucket on the 190-4, essential when working in close quarters or when loading trucks on single traffic lanes in busy streets. After purchasing the Dynahoe line in 1985, Bucyrus Construction Products continued to build the 190-4 until new models were introduced.

Ruston-Bucyrus 190-4. The popular 4-wheel-drive 190-4 loader-backhoe was distributed in the United Kingdom by Ruston-Bucyrus Ltd. beginning in 1974. Sold as an R-B machine, its specification was identical to the American model except for a Ford 2715E diesel engine developing 108 horsepower to replace the Detroit engine. Sales were discontinued in 1979, likely as a result of stiff competition from lower-priced, home-built machines.

Dynahoe 200-4. This 20-foot digging-depth model was introduced in 1968 by Hy-Dynamic Company. It was the company's largest machine, and its first 4-wheel drive model. Like the smaller 190-4 it was provided with rear-wheel steering. Two years after Bucyrus-Erie acquired the Dynahoe manufacturing rights in 1971 it closed the Lake Bluff, Illinois, plant and transferred Dynahoe manufacture to existing B-E plants at Evansville, Erie and South Milwaukee.

Dynahoe 200-4B. The former Dynahoe 200-4 became the 200-4B under the Bucyrus-Erie banner. With an operating weight of 38,490 pounds including ROPS cab, it was the largest Dynahoe model. It carried a GM Detroit 4-71 diesel engine of 163 horsepower and could achieve a top road speed of 18 mph. The loader bucket carried 2 1/4 cubic yards, and the backhoe could pick up a load of 11,600 pounds.

Dynahoe 1900. The 1900 was the final Dynahoe model Bucyrus-Erie launched before it sold the line in 1985. Making an appearance in 1983 it was a rework of the former model 190. Digging depth was the same 19 feet but its Detroit 3-53 diesel was now turbocharged with horsepower upped to 125. A weight increase to 23,250 pounds with ROPS cab, loader bucket increase to 1 3/4 yards, hoe lifting capacity to 8,800 pounds, and a hydraulic pressure increase from 2,150 to 3,000 psi completed the picture.

Dynahoe 220E. The Hy-Dynamic Company was developing a fully revolving wheeled hydraulic excavator just about the time Bucyrus-Erie took over the company. The 220E, bearing similarities to the revolving rough terrain cranes already built by Hy-Dynamic, continued to be developed, and was marketed by Bucyrus-Erie for a short time up to about 1974. It was ahead of its time as wheeled excavators had not yet become popular in the United States.

Chapter 12
Cranes

The tremendous variety of cranes produced by the Bucyrus companies originated in the 1800s to satisfy the needs of its railroad shovel customers. A few made-to-order rail-mounted cranes were produced before the turn of the century, including some standard railroad shovels fitted with crane booms, and an electric-powered model in 1895. The first true Bucyrus railroad "wrecking" cranes appeared circa 1900. In 1905 one 100-ton and two 75-ton units were sold to the Isthmian Canal Commission for use on the Panama Canal railroad, a huge undertaking preceding the canal itself. Railroad "wrecking" cranes formed a substantial part of Bucyrus' products in the first half of the 20th Century, ranging from 5 to 250 tons capacity. By 1956 some 230 had been built and Bucyrus-Erie finally announced withdrawal from this market in 1963.

When universal excavators became popular in the 1920s it was a simple matter to replace the dragline bucket with a hook block, and the crawler crane was born. For greater lifting capacity more counterweight was added, then long and wide undercarriages were introduced to boost lifting even further. This concept reached its peak in the 1970s when Ruston-Bucyrus in the U.K. introduced its SC-series "Super Cranes" with lifting capacities reaching to 135 tons in its largest model, the 1200-SC.

Early truck cranes were also derived from universal excavators when they were mounted on roadable truck chassis. Many of Bucyrus-Erie's universal excavators were sold optionally as truck-mounted versions, which turned them into mobile cranes when provided with crane booms. By the early 1960s truck cranes had become sophisticated, built-for-the-purpose mobile units with lighter booms claiming to "lift more, higher up, further out." Hydraulics took over with the telescopic boom concept, and Bucyrus-Erie competed in this market with an extensive line of hydraulic rough terrain and truck- mounted models.

Bucyrus also built various types of specialized cranes over the decades, often purchasing the rights from smaller companies. In 1931 it acquired the manufacturing rights of the "Loadmaster" crane from Frederick H. Poor Company of New York. The Loadmaster crane, offered in several sizes, was a fully revolving crane that mounted on top of a crawler tractor. In 1940 Bucyrus-Erie built cranes under license from Cullen-Friestadt Company of Chicago. The undercarriage of these cranes was equipped with both crawler tracks and flanged rail wheels. The crawler tracks were vertically adjustable so the machine could travel on railroad track or on the ground. In 1948 the Milwaukee Hydraulics Corporation was acquired with its Hydrocrane, a hydraulically operated truck-mounted crane that evolved into the Hydrohoe excavator. (See Chapter 10)

In 1960 Bucyrus-Erie commenced manufacture of a tower crane, the first if its type designed and built in America. In 1963 Ruston-Bucyrus in the U.K. purchased a mobile tower crane design from Star Cranes Ltd. fitted with wheels for towing on the highway and crawler tracks for mobility on site. Bucyrus-Erie responded to the oil boom of the 1970s with lines of ship cranes and marine cranes for service on offshore oil rigs.

As with most construction equipment manufacturers, Bucyrus-Erie experienced a drastic decrease in demand for its machines in the early 1980s and, by 1985 the company had sold or discontinued most of its construction-related lines and transformed itself to a company serving only the surface mining industry.

15-Ton Derrick. Solving the needs of its early railroad customers brought Bucyrus into crane manufacture. This early car-mounted derrick with chain hoist affixed to the boom is an example shown in an 1898 catalog. It boasted 15 tons capacity with no motive power and its hoist and swing motions controlled by hand cranks. Heavy iron castings and riveted construction mark the 1880s era.

Rail Crane. In 1895 Bucyrus built its first electric crane. Rated at 1-ton capacity it was equipped with a chain hoist and stabilizers set by hand wheels. Although this special crane was designed for some specific application lost in history, its design is similar to other steam-powered cranes of the era, with capacities up to 5 tons.

60-Ton Wrecking Crane. This early steam railroad "wrecking" crane built around 1900 set the Bucyrus design for decades to come, though little operator comfort or visibility is evident. Equipped with a double-bogie car and six heavy-duty outriggers, its capacity reached 60 tons. In 1907 it was upgraded to a similar 75-ton model.

Class 150-17 Railway Crane. Bucyrus applied this nomenclature on most of its railroad cranes after the formative years. It indicated the crane's maximum capacity at minimum radius. Thus the steam crane illustrated could lift 150 tons at 17 feet radius. Offered in propelling or non-propelling versions, it weighed approximately 125 tons in operation. Forty-nine of this model were built from 1918 to 1929.

250-Ton Railway Crane. Shown under test is a 250-ton diesel-electric railway crane, one of two delivered to Quebec North Shore & Labrador Railway Company in 1953. Bucyrus introduced the steam version of this model, its largest railway crane, in 1942 with air control, and the diesel-electric version in 1948 with Ward-Leonard control. The main hoist could lift 250 tons, the auxiliary hoist 60 tons and the whip line 15 tons.

Loadmaster CT-50. The Loadmaster fully revolving tractor crane originating with Frederick H. Poor Company was acquired by Bucyrus-Erie in 1931. Mounted on any suitable tractor, the initial model had a maximum lifting capacity of 3,500 pounds. By 1934 Bucyrus offered a range of models with capacities from 5,000 to 13,900 pounds to fit both wheel and crawler tractors. The largest, the CT-50 illustrated, was based on a Caterpillar Diesel Fifty crawler tractor.

60 Loadmaster. An upgraded version of the CT-50 appeared in 1936. With design and capacity similar to the former model, the 60 Loadmaster was based on the Caterpillar RD7 diesel tractor rated at 70 horsepower. The tractor undercarriage was modified to incorporate hydraulic stabilizers to hold the machine rigid during lifting. Standard boom length was 30 feet and the entire outfit weighed just over 18 tons.

H-3 Hydrocrane. Bucyrus-Erie's first move toward hydraulic excavators came in 1948 when it purchased the Milwaukee Hydraulics Corporation and its "Hydrocrane" truck-mounted hydraulic crane. After some redesign, the H-3 was born which could lift 3 tons and extend its boom from 18 to 26 feet. A pair of hydraulic cylinders raised the boom and another set of cylinders working between two sets of sheave blocks provided the hoist. The 5-ton capacity H-5 appeared in 1955.

Mark I-50 Tower Crane. Bucyrus-Erie designed and built one of the first all-American tower cranes in 1960. The Mark I-50 featured a triangular tower with luffing jib, and counterweight sections positioned on the revolving base. The I-50 traveled on a 16-foot gauge track, could self-erect without assistance, and was electrically powered. At maximum 120-foot radius, 5,300 pounds could be lifted to a height of 155 feet.

37-B Wagon. During the 1930s and 1940s before the advent of cheaper single-purpose mobile cranes, many of Bucyrus-Erie's universal excavators found use in industrial and military applications as wagon cranes. This example, purchased by the U.S. Marine Corps as their fleet number 32, is a 37-B 30-ton crane with 70-foot boom mounted on a Maxi BHD 4-16 undercarriage.

34-80 Tower Crane. When tower cranes became popular in the 1960s, Ruston-Bucyrus purchased the manufacturing rights to a portable tower crane designed by Star Cranes Ltd. of Horsham, England. It was towed to the site on its own wheels, erected itself, and then used crawler tracks for mobility. The initial 1963 model **12-52** could lift 1,344 pounds at maximum 52-foot radius with 52-foot height under the hook. The larger 34-80, introduced in 1966, lifted 3,920 pounds at 80-foot radius. Tower cranes were phased out after 1970.

25-Ton Transit. When the Series III 22-B excavator was introduced, truck crane versions were also introduced, simply known as the 25-Ton and the 30-Ton Transit cranes rated with those stated capacities. Their almost-identical 6 x 4 carriers were specially built for Bucyrus-Erie and powered by an International RD-501 gasoline engine. The 25-Ton Series III combined unit weighed just over 30 tons.

22-B Transit. Most of Bucyrus-Erie's universal excavators up to the 30-B could be supplied truck-mounted. One of the most popular was the 22-B which first appeared in 1941. Mounted on a 6 x 4 carrier specially built for crane duty, it was initially rated at 20 tons 10 feet radius. By 1954 this rating had increased to 25 tons. Subsequent upgrades to 30 tons capacity led to the T-series of truck cranes in which the 22-B was the base machine for the 25-ton **25-T**, 35-ton **35-T**, and 40-ton **40-T**.

50-T. The 50-ton capacity 50-T transit crane was based on the air-controlled 30-B Series III excavator. The upper crane ran with its normal diesel engine but was mounted on an 8 x 4 carrier driven by a Waukesha 238-horsepower gasoline engine. The 60-ton **60-T** followed in 1966, and the 70-ton **70-T**, based on the 30-B Series V, appeared in 1981 with carrier powered by a GM 6-71T diesel rated at 262 horsepower.

400SC. Ruston-Bucyrus launched a broader line of "Super" crawler cranes beginning in 1975. They were really upgraded crane versions of standard R-B excavators with more counterweight, long and wide undercarriages, and special crane features such as removable crawler side frames and torque converter drive. The 55-ton 400SC and the 62-ton capacity **550SC** were super versions of the 30-RB, the latter with 16-foot 9-inch long by 13-foot 6-inch wide undercarriage.

110-T. The 110-T was built from the ground up as a 110-ton capacity truck crane. Introduced in 1964 it was a revision of the earlier 100-T appearing a couple of years earlier. It featured the new Bucyrus-Erie "Trigon" triangular boom, and its 8 x 4 carrier was designed and built by Bucyrus-Erie. With its lightweight boom and balanced design, advertisements boasted the 110-T could lift 69% more than its own weight.

800SC. The 61-RB was the base machine for Super Cranes 800SC, **1100SC** and **1200SC** with maximum lifting capacities of 90, 125 and 136 tons respectively. As with the 61-RB these machines featured a double-row ball-bearing swing circle. To round out the bottom end of the SC-series cranes, the 22-RB was upgraded to the 300SC in 1977 with a lifting capacity of 33 tons. 350SC and 600SC models were added in 1980 and 1983.

45-C. Bucyrus-Erie offered a line of in-house designed hydraulic truck cranes beginning with the 30-ton 45-C in 1967. Available with 6 x 4 or 8 x 4 chassis and remote-controlled outriggers, the machine could be ready for work in five minutes after arrival at site. The boom telescoped to 85 feet and a 20-foot jib section, carried on the machine, could also be added.

65-C. Three more hydraulic truck cranes were added in 1968: the telescoping 20-ton 35-C, and two lattice-boom models, the 45-ton 55-C, and 70-ton 65-C. The latter two embodied hydraulically driven hoist drums through precision planetary units. The 65-C could carry 180 feet of boom plus a 50-foot fly jib. These hydraulic cranes were branded with the Hydrocrane name.

60-XC. In 1977 Bucyrus-Erie upgraded its hydraulic transit cranes and launched the XC-series of telescopic boom cranes. The four new models were the **20-XC, 25-XC, 30-XC** and 60-XC, their model numbers indicating rated capacity. They were structurally designed to reduce machine weight. The 60-XC with 60-ton capacity weighed only 38 tons with its 104-foot boom and 32-foot fly jib. The 25-XC, 35-XC and 45-XC were built in the U.K. by Ruston-Bucyrus.

90-XC. The XC-series was rounded out at the top end with the 90-ton 90-XC in 1979. Mounted on a 10 x 4 chassis, the 90-XC was equipped with a main telescoping boom 144 feet long, and jibs up to 54 feet could be added. Bucyrus-Erie's Hydrocrane line of hydraulic transit cranes, including the original model H-3, survived until 1981 when they were discontinued.

200-C. Bucyrus-Erie purchased the Hy-Dynamic Company of Lake Bluff, Illinois, in 1971. Well-known for its Dynahoe loader-backhoes (see Chapter 11), it also presented a line of rough-terrain "cab-down" cranes known as the C-series, with model designations indicating lift capacities in pounds. The illustration shows a prototype 200-C of 10-ton capacity. The line eventually extended up to the 60-ton **1200-C** launched in 1978.

360-C. The 18-ton 360-C was the top-of-the-line rough terrain crane from its introduction in 1971 until the 1200-C appeared in 1978. These cranes also carried the brand name "Hydrocrane" although having a different pedigree from its stable-mate XC-series truck cranes bearing the same name. They featured a ball-bearing swing circle and four-wheel-steer chassis. Soon after 1980 all the Hydrocrane products were phased out.

MK-100. Many of Bucyrus-Erie's standard model excavators and cranes were available pedestal mounted for use on ships, docks and offshore oil rigs. In addition, a line of fully hydraulic single-purpose marine cranes was presented to the industry in 1970. The initial three models, **MK-35**, **MK-60** and MK-100, were rated at 31,000, 60,000 and 115,000 pounds lifting capacity at 50-foot radius with 120-foot booms. In 1997 Bucyrus sold its marine crane product line to Offshore Crane Engineering Ltd. of Abedeen, Scotland.

20-Ton Cargo Crane. This is an example of one of Bucyrus-Erie's many special-duty cranes built to satisfy specific needs. In this case a telescopic hydraulic crane has been adapted with interesting geometry to serve as a ship's cargo crane. Others examples included high-cab dockside cranes, hydraulic winches, and product lines of cable and hydraulic logging cranes utilizing heel booms.

Chapter 13
Continuous Excavators

Where geological conditions allow, the continuous excavator can deliver the highest output of any digging machine. Continuous excavators include any machine that excavates and dumps material in a continuous stream as opposed to the cyclic action of shovels and draglines. The Bucyrus-Erie built machines in this category were trench excavators and bucket wheel excavators.

Bucyrus built its first continuous excavator, a ladder-type ditcher, as far back as 1915. Then in 1939 it commenced building a machine called the Ruth Dredger, a very adaptable machine used not only for trenching but also in a wide variety of canal excavation and cleaning projects. This machine was well established, having been built since 1906 by the Ruth Dredger Corporation of Los Angeles, California.

In 1954 Bucyrus-Erie placed in service its first bucket wheel excavator. It was the model 954-WX purchased by Truax-Traer Coal Company for their coal mine at Fiatt, Illinois. Most of the wheel machines Bucyrus built were known as "cross-pit" bucket wheel excavators. They were all "one-off" machines, each built for a particular mining operation. They were also integrated units designed with large dimensions, enabling them to dig from the high wall of a surface coal mine and discharge clear across the pit onto the spoil pile in one continuous operation. The digging wheel, transfer conveyor, machinery house and discharge conveyor were part of a revolving superstructure mounted on a crawler-propelled base, often from a former stripping shovel. The nomenclature of the Bucyrus-Erie wheel models indicated the size of the shovel base in the first digit, followed by the width of the discharge belt in inches. Thus the 954-WX was based on a 950-B shovel base and utilized a 54-inch belt.

In 1959 Bucyrus-Erie enhanced its bucket wheel capabilities by purchasing the plans and patents of the bucket wheel excavators designed and built by the United Electric Coal Company of Chicago. This company, spearheaded by its president Frank W. Kolbe, had built several of these machines for use in its own coal mines in Illinois as far back as 1944, and had already licensed Bucyrus-Erie to build the 954-WX five years earlier.

Bucyrus Class 72. Bucyrus built its only model trench excavator for a few years beginning in 1915. The steam-powered Class 72 weighed some 40 tons, and could dig 17 feet deep with a standard ladder or 20 feet deep with optional extension. It dug trenches from 24 to 72 inches wide, the higher number indicating the Class 72 model number. The conveyor was driven by an independent steam engine and the excavated material could be discharged on either side.

Bucyrus-Ruth HU. In 1939, Bucyrus-Erie acquired the rights to manufacture a unique ditching machine designed by the Ruth Dredger Manufacturing Corporation which was established in 1906. The Bucyrus-Ruth machine could swing its boom with buckets laterally through 180 degrees, so riverbanks could be trimmed on either side of the machine as it moved forward. The model HU carried a 32-horsepower Caterpillar D4400 engine and weighed 18 tons. This product was discontinued in 1942.

Kolbe W3A. In 1981 Freeman United Coal Mining Company, formerly the United Electric Coal Company, rebuilt one of its five cross-pit bucket wheels for use at its new Industry Mine, Illinois. Originally mounted on an old Marion 5480 stripping shovel, it was remounted on the base of an old 1050-B stripping shovel. The company claimed the wheel excavator produced up to five times more than the former shovel mounted on the same crawler undercarriage. *Photo courtesy of Urs Peyer*

684-WX. Bucyrus-Erie built this wheel excavator for Morrison-Knudsen, contractors on the 80-million yard San Luis Dam project in California. Starting work in 1964, the 812-ton machine was designed to load trucks at 4,000 cubic yards per hour using its 30-foot diameter wheel carrying ten 2 1/2-yard buckets. Mounted on a two-crawler 280-B shovel-type undercarriage, the 684-WX was electrically powered.

684-WX. After successfully completing the California dam job, the 684-WX was dismantled and shipped to Alberta, Canada, where it spent over 20 years excavating at the Suncor oil sands operation. The 84-inch conveyor could load a 100-ton truck in 45 seconds. Now retired, the 684-WX is a key exhibit preserved at the Oilsand Interpretive Center at Fort McMurray, Alberta.

954-WX. This was the first bucket wheel excavator built by Bucyrus-Erie and put into service in 1954 by Truax-Traer Coal Company of Chicago. Although its undercarriage was based on a 950-B stripping shovel, the machine was designed as a wheel excavator from the ground up. Its 24-foot diameter wheel carried nine buckets, each holding one cubic yard. Digging height was 75 feet above crawlers, and material was moved 275 feet from wheel to discharge point.

1054-WX. Built on a retired Bucyrus-Erie 1050-B 8-crawler stripping shovel base, the 1054-WX weighed approximately 1,600 tons and could cut a face 100 feet high. The excavated material traveled over 370 feet from wheel to the spoil pile. Two 1,000-horsepower AC motor generator sets powered DC motors for the machine's various functions. The 1054-WX started work in 1960 at Peabody Coal Company's River King Mine, Illinois.

1060-WX. Ordered by Consolidation Coal Company in 1967 for its Burning Star No. 2 Mine, Illinois, the 1060-WX boasted an overall length from bucket to end of discharge conveyor of 454 feet, or one-and-a-half football fields. The Bucyrus-Erie cross-pit bucket wheels incorporated a crowding action where the wheel boom could be forced into the bank under power. Crowding of 51 feet was possible on the 1060-WX.

1060-WX Lower. The 1060-WX was built from scratch as a totally new machine. Its undercarriage consisted of four 2-crawler assemblies each supporting a hydraulic leveling cylinder like the stripping shovels. The crawlers supporting the 1,735-ton machine weight exerted a ground pressure of 49 psi. The various electric motors on this machine totaled 2,775 horsepower.

5872-WX. Bucyrus-Erie built the world's largest cross-pit wheel excavator and placed it in service at the Captain Mine of Arch Coal Inc., Illinois, in 1986. The 5872-WX rotated a 40-foot diameter wheel with 12 buckets, each of 2.14-cubic-yards capacity. Excavated material moved from the digging face to the spoil pile, a distance of more than 700 feet!

5872-WX. The 5872-WX wheel at the Captain Mine was built on the lower works of a retired 80-yard Marion 5860 stripping shovel. Overall the machine's estimated weight was some 5,380 tons. The height of the supporting mast reached up 250 feet and its 72-inch discharge belt moved at 1,200 feet per minute. After the Captain Mine closed, the 5872-WX was scrapped in 1999. *Photo courtesy Eric C. Orlemann*

Chapter 14
Miscellaneous Machines

This final chapter covers the Bucyrus machines that don't fit into any of the main categories previously dealt with. While related to Bucyrus' main products and meeting a customer's preference for purchasing from a known supplier, these special products were built in relatively small numbers. Bucyrus built railway spreader plows, snowplows and locomotive pile drivers, as well as ballast unloaders (not illustrated) to satisfy its railroad shovel customers. And the tower excavators, car transporters and Chambers Bridge were designed to solve particular problems encountered by its dragline and shovel customers. In addition Bucyrus proposed many other machines that, for various reasons, did not advance beyond the design stage—a reflection of a company always striving to experiment with new ideas and bring the latest technology to its customers.

Class 50 Spreader Plow. Working like a rail-mounted motor grader, the spreader plow was pushed by a locomotive. Its function was to spread material dumped by rail cars, forming berms or banks in readiness for the next trainload. The 70-ton spreader plow was used not only for railroad construction and maintenance, but also in large mining operations to build high dumps before the age of crawler bulldozers. Blade motion was air-operated from a steam-driven air compressor.

Locomotive Pile Driver. As its name suggests the Bucyrus pile driver was self-propelled and able to move itself plus a 300-ton work train at speeds of up to 15 mph. It was designed for driving piles with either steam or drop hammers at distances up to 18 feet ahead of the forward wheels. The machine shown is fitted with an optional turntable consisting of a swing bearing and carbody that could raise the entire machine by hydraulic (water) cylinders and enable piling within a 32-foot diameter circle. Locomotive pile drivers were built from 1902 to 1916.

Rotary Snow Plow. The rotary snow plow was essential for keeping northern railroads open in winter months. Designed to be pushed ahead of a locomotive, it embodied its own steam boiler and a double 18-inch by 26-inch bore and stroke steam engine. The rotary knives were driven at high speed through heavy-duty shaft and steel bevel gearing.

Bucyrus 18 x 26 Rotary Snowplow—built for the Grand Trunk Pacific Railway Company.

101

Car Transporter. To take advantage of a shovel's faster digging cycle, while at the same time providing the range comparable to a dragline, Bucyrus-Erie developed the crawler-mounted car transporter for surface mining in 1932. The car could be conveyed up 45-degree inclines, and was timed to dump and return in unison with the shovel's cycle. Any size of material handled by a shovel dipper could be carried, without the need for crushing as in methods utilizing conveyors.

Chambers Bridge. Conceived by John T. Chambers of Lewis-Chambers Construction Company, and built by Bucyrus-Erie commencing in 1934, the machine known as the Chambers Bridge was used in wide canal excavation and Mississippi drainage projects. The machine was mounted on a crawler-mounted frame with a tower supporting booms at each side. It operated on similar principles to the car transporter except with far greater range and capacity. The 10-yard car traveled a distance of 350 feet and could discharge at any point along its track.

TX-310 Tower Excavator. A close cousin to the dragline, the tower excavator was introduced in 1915. Originally designed for constructing levees along the lower Mississippi River, it later found use in open pit mines. The tower excavator consisted of a head tower containing the draw works and a tail tower positioned at the opposite side of the excavation. A dragline bucket, running on a trolley between the two, dragged the excavated material into a hopper near the head end.

Tower Excavator Tail Tower. This is a typical tail tower installation matching the machine in the previous illustration. Both are mounted on 8-crawler undercarriages for maximum mobility, but on more static operations such as open pits, either the tail tower or the head tower could be fixed. Bucyrus' largest model of this type was the TX-310 which carried a dragline bucket of 10-cubic-yards capacity over a span of 750 feet. The head tower was electrically powered but a diesel powerplant propelled the tail tower.

Walking Tail Tower. With the Bucyrus-Monighan walking system figuring so prominently in the company's draglines, it was only natural that the same propel system would be applied to a tower excavator tail tower. Under full stride this machine is being relocated on a U.S. Army Corps of Engineers job at Slough Landing Neck, Kentucky, in 1932.

Dump Trailer. As a special project, Bucyrus-Erie was commissioned to build a fleet of 24 side-dumping trailers for the Heil Company of Milwaukee in 1962. The trailers, each with a capacity of 100 tons, were built at the Guelph, Ontario, plant of Bucyrus-Erie Company of Canada Ltd. They worked at the Iron Ore Company of Canada's Carol Lake mine in Labrador where they were hauled by Euclid 40-ton prime movers.

150-B Doubler. The "Doubler" was so named because it theoretically doubled the capacity of the electric mining shovel to which it was fitted. Designed by Heckett Engineering, Illinois, and built to compete with Marion's Superfront, the Bucyrus-Erie machine employed the shovel's regular electric system in conjunction with a hydraulic system to provide dipper hoist and wrist action. Only three were built from 1968 to 1970, converted from two 150-B and one 190-B shovels.

Cable Carrier. The operation of larger electrically powered excavators requires precise and careful handling of their high-cost power cables. When long lengths are required to reach from the power supply to the machine, a power-operated cable carrier can pay for itself in prolonging cable life. This sophisticated crawler-mounted model is ready to serve a large walking dragline.

Howitzer Gun Carrier. World War II brought lots of business for Bucyrus-Erie, mostly for its standard excavator and tractor products. However, as in World War I, the company did execute several orders for artillery equipment including carriers for this 240mm howitzer gun. Posing with the unit is a group of senior engineers, President N.R. Knox (third from right), and Chairman W.W. Coleman (second from right).

APPENDIX 1

(A) CABLE EXCAVATORS PRODUCTION AND SHIPPING DATES

Including mining shovels, stripping shovels, non-walking draglines and Ruston-Bucyrus (R-B) machines

Model	*Standard Dipper c.y.*	*Manufactured From*	*To*	*Mounting C R W S**	*Power S G D E A**	*Original Design by*	*Number Built*	*Notes*
AA	1/2	1929	1935	C	A	B-E	n/a	Special Tunnel Shovel
A	1/2	1916	1931	C W	S A	Erie	531	
B	3/4	1914	1925	C R W	S	Erie	3,130	
B-2	7/8	1925	1930	C R W	S A	Erie	562	
B-3	7/8	1929	1936	C R	S A	Erie	n/a	
GA-2	1	1925	1930	C	G+A	Erie	451	
GA-3	1 1/4	1930	1935	C	G+A	Erie	38	
DA-2	1	1927	1930	C	D+A	Erie	29	
DA-3	1	1930	1931	C	D+A	Erie	n/a	
D-2	1	1927	1933	C	D	Bucyrus	175	
E-2	1 1/4	1927	1930	C	D	Bucyrus	57	
TS-5	1 1/4	1931	1934	R	A	B-E	3	Special Tunnel Shovel
1020	1/2	1929	1934	C	G D E	Erie	81	
1020 (R-B)	1/2	1930	1931	C	G	Erie	2	
1025	5/8	1930	1932	C	G E	Erie	7	
1025 (R-B)	5/8	1931	1932	C	G	Erie	2	
1030	3/4	1928	1933	C	G D E	Erie	227	
1030 (R-B)	3/4	1930	1932	C	G D E	Erie	65	
1035	7/8	1929	1931	C	G E	Erie	63	
1060		Ca.1929		C	G	Erie	1	
2030	5/8	1929	1929	C	G	Erie	33	Built for Pennsylvania Railroad
7 Class	1	1915	1923	C R W S	S G E	Bucyrus	n/a	Dragline only
9 1/2 Class	1 1/2	1913	1925	C S	S G E	Bucyrus	n/a	Dragline only
10-B	3/8	1934	1959	C	G D E	B-E	2,428	
10-RB	3/8	1934	1969	C	G D E	B-E	7,625	
11-B	3/8	1958	1960	C T	G D	B-E	412	
11-B II	3/8	1960	1964	T	G D	B-E	Incl. above	
11-RB	3/8	1961	1963	T	G D	B-E	35	
14 Class	2	1912	1930	4C S	S G D E	Bucyrus	n/a	Dragline only
14-B (Steam)	5/8	1912	1922	C R W	S E	Bucyrus	188	
14-B	1/2	1960	1967	C R T	G D	B-E	240	
14-RB	1/2	1934	1937	C	G D E	B-E	115	Same as 16-RB except power.
15-B	1/2	1940	1961	C T	G D E	B-E	4,099	
15-B II	5/8	1961	1968	C T	G D E	B-E	Incl. above	
16-B	1/2	1932	1939	C	G D E	B-E	76	
16-RB	5/8	1934	1937	C	G D E	B-E	155	
17-B	1/2	1936	1938	C	G D E	B-E	21	
17-RB	1/2	1937	1942	C	G D E	B-E	264	
18-B (Steam)	7/8	1912	1920	C R W	S E	Bucyrus	171	
18-B	5/8	1938	1942	C	G D E	B-E	35	
19-B	5/8	1936	1938	C	G D E	B-E	120	
19-RB	5/8	1937	1967	C	G D E	B-E	5,437	
20 Class	2 1/2	1912	1924	R S	S E	Bucyrus	n/a	Dragline only
20-B (Steam)	3/4	1922	1928	C R W	S G D E A	Bucyrus	413	
20-B	3/4	1938	1943	C	G D E	B-E	582	
20-RB	3/4	1939	1942	C	G D E	B-E	115	
21-B	3/4	1931	1937	C	G D E	B-E	41	
21-RB	3/4	1932	1939	C	G D E	B-E	156	
22-B	3/4	1937	1958	C	G D E	B-E	8,828	
22-B II	3/4	1960	1964	C T	G D E	B-E	Incl. above	

Model	Standard Dipper c.y.	Manufactured From	To	Mounting C R W S*	Power S G D E A*	Original Design by	Number Built	Notes
22-B III	3/4	1964	1976	C	D E	B-E	Incl. above	
22-RB	3/4	1950	1995	C T	G D E	B-E	10,245	Includes all variations and models.
24 Class	3 1/2	1911	1930	C R S	S D E	Bucyrus	n/a	Dragline only
24-RB	7/8	1938	1955	C	G D E	B-E	506	Made by R-B at Lincoln only
25-B (Steam)	1 1/4	1914	1920	R	S	Bucyrus	12	
25-B	1	1961	1964	C	G D	B-E	807	
25-B II	1	1964	1968	C	G D	B-E	Incl. above	
25-B III	1	1968	1984	C	G D	B-E	Incl. above	
27-D	1 1/2	1915	1922	C	S	Bucyrus	31	Skimmer only
29-B	1	1937	1942	C	G D E	B-E	44	
30-B (Steam)	1	1920	1928	C R W	S G D E	Bucyrus	328	
30-B	1	1956	1960	C T	G D E	B-E	3003	
30-B II	1 1/4	1960	1964	C T	G D E	B-E	Incl. above	30-RB II made 1968-1971
30-B III	1 1/4	1964	1968	C	D E	B-E	Incl. above	
30-B IV	1 1/4	1968	1981	C	D E	B-E	Incl. above	30-RB IV made 1971-1982
30-B V	1 ½	1981	1984	C	D E	B-E	Incl. above	30-RB V made 1982-1988
30-RB	1	1956	1988	C	D E	B-E	2,135	Includes Series I to V
30-RB VI	1 ½	1988	1995	C	D	R-B	Incl. above	Made by R-B at Lincoln only
31-B	1	1926	1928	C	S A	Bucyrus	85	
32-B	1	1931	1937	C	G D E	B-E	51	
32-RB	1	1932	1938	C	G D E	B-E	117	
33-B	1 1/4	1937	1942	C	G D E	B-E	96	
33-RB	1 1/4	1938	1956	V	G D E	B-E	527	
34-B	1 1/4	1933	1937	C	G D E	B-E	49	
35-B	1 1/2	1914	1922	C R W	S E	Bucyrus	95	
37-B	1 1/2	1931	1943	C	G D E	B-E	346	
37-RB	1 1/2	1932	1949	C	G D E	B-E	270	
38-B	1 1/2	1945	1962	C	G D E	B-E	2,188	38-RB made 1948-1967
38-B II	2	1963	1983	C	D E	B-E	Incl. above	38-RB II made 1965-1988
38-RB	1 1/2	1948	1988	C	D E	B-E	1,811	Includes all variations and models.
41-B	1 1/4	1926	1930	C	S E	Bucyrus	110	
42-B	1 3/4	1930	1936	C	S E	B-E	37	
43-B	1 3/4	1931	1939	C	G D E	B-E	115	43-RB made 1931-1955
43-RB	1 3/4	1931	1955	C	G D E	B-E	287	
44-B	2	1939	1945	C	G D E	B-E	122	
45-B	2	1932	1933	C	G D	B-E	n/a	Dragline only
46-B	2	1933	1935	C	G D	B-E	n/a	Dragline only
48-B	2 1/4	1936	1939	C	G D E	B-E	20	
48-RB	2	1938	1939	C	E	B-E	4	
50-B	1 3/4	1922	1934	C R	S G D E	Bucyrus	534	
51-B	2	1946	1959	C	G D E	B-E	333	
51-B II	2 1/2	1960	1964	C	D E	B-E	Incl. above	
52-B	2 1/4	1930	1939	C	D E	B-E	17	
52-RB	2 1/4	1931	1936	C	D E	B-E	26	
54-B	2 1/2	1939	1959	C	D E	B-E	997	
54-B II	3	1960	1964	C	D E	B-E	Incl. above	
54-RB	2 1/2	1943	1973	C	D E	B-E	586	
55-B	2 1/2	1935	1939	C	D	B-E	7	
55-RB	2 1/2	1936	1945	C	D E	B-E	64	
61-B	3	1963	1967	C	D	B-E	348	
61-B II	3 1/4	1967	1984	C	D	B-E	Incl. above	
61-RB	3	1967	1990	C	D	B-E	179	
62-B II		1975	1984	C	D	B-E	n/a	Slag crane, made for overseas order
65-D	3	1976	1979	C	D	B-E	100	Clamshell + Dragline Only
65-D II	3	1979	1981	C	D	B-E	Incl. above	Clamshell + Dragline Only
65-D III	3	1981	1983	C	D	B-E	Incl. above	Clamshell + Dragline Only

Model	Standard Dipper c.y.	Manufactured From	To	Mounting C R W S*	Power S G D E A*	Original Design by	Number Built	Notes
71-B	3	1954	1959	C	D E	B-E	641	
71-B II	3 1/2	1960	1967	C	D E	B-E	Incl. above	
71-B III	4 1/4	1967	1983	C	D E	B-E	Incl. above	71-RB III Made 1971-1992
71-RB	3 1/2	1961	1992	C	D E	B-E	229	Includes all variations and models.
72-B		1959	Ca.1977	C	E	B-E	23+	Special 3-motor electric slag crane
73-B		1964	1974	C	E	B-E	11+	Single motor electric slag crane
75-B	2 1/4	1928	1935	C	E	B-E	28	
80-B	2 1/2	1921	1929	4C R	S E	Bucyrus	n/a	Dragline version: "Class 80"
84-B	4 1/2	1964	n/a	C	D E	B-E	n/a	Special 88-B
85-B	3	1935	1957	C	E	B-E	71	
87-D	3	1917	1917	4C	S	Bucyrus	n/a	4-crawler skimmer, one made?
88-B	4	1946	1959	C	D E	B-E	617	
88-B II	5	1960	1962	C	D E	B-E	Incl. above	
88-B III	5	1962	1968	C	D E	B-E	Incl. above	
88-B IV	5 1/2	1968	1985	C	D E	B-E	Incl. above	
88-B Loader	10	1981	1984	C	D E	B-E	Incl. above	
100-B	3 1/2	1926	1950	C	S E	Bucyrus	150	
100-RB	3 1/2	1933	1954	C	E	B-E	83	
110-B	4 1/2	1950	1975	C	D E	B-E	92	
110-RB	4 1/2	1955	1978	C	E	B-E	151	
111 Class	3	1931	1939	C	D	B-E	n/a	Special dragline only
114-B	6	1970	1970	C	E	B-E	1	Special dragline only
115-B	6	1962	1964	C	D E	B-E	3	Special dragline/Crane only
120-B	5	1925	1951	C	S E	Bucyrus	303	
120-RB	5	1948	1954	C	E	B-E	30	
150-B (Old)	2 1/2	1911	1924	R	S	Bucyrus	17+	Stripping shovel
150-B	6	1951	1975	C	E	B-E	151	
150-RB	6	1957	1983	C	E	B-E	125	
155-B	9	1975	1980	C	E	B-E	10	
155-B I	9	1982	1982	C	E	B-E	1	
170-B	6 1/2	1929	1951	C	E	B-E	84	
175-B	3 1/2	1912	1925	R	S E	Bucyrus	54+	Stripping shovel. Dragline: "Class 175"
180 Class	12	1915	1920	R	S	Bucyrus	n/a	
182M	12	1999	Current	C	E	Marion	n/a	Former Marion 182-M
190-B	9	1952	1976	C	D E	B-E	118	
195-B	13	1968	1980	C	E	B-E	43	
195-B (RB)	12	1974	1985	C	E	B-E	44	
195-KB	13	1976	1976	C	E	K-B	2	Made by Komatsu-Bucyrus in Japan
195-B I	13	1978	1998	C	E	B-E	40	
200-B (Old)	7	1927	1943	8C R	S E	Bucyrus	20	Stripping shovel. Dragline: "Class 200"
200-B	16	1970	1978	C	E	B-E	5	Special dragline only
210-B	9	1961	1966	C	D E	B-E	6	Special dragline only
225-B	6	1914	1923	R	S E	Bucyrus	91+	Stripping shovel. Dragline: "Class 225"
230 Class	6	1917	1931	8C R	S D	Bucyrus	n/a	
270-B	9	1960	1962	C	E	B-E	5	
280-B	15	1962	1982	C	E	B-E	81	
280-KB	15	1971	1976	C	E	B-E	14	Made by Komatsu-Bucyrus in Japan
290-B I	17	1979	1982	C	E	B-E	14	
290-B II	17	1982	1987	C	E	B-E	3	
295-B	21	1972	1980	C	E	B-E	48	
295-B I	27	1980	1981	C	E	B-E	7	
295-B II	27	1982	1998	C	D E	B-E	38	
295-B III	27	1993	1996	C	E	B-E	6	New 295-series is current
300-D	12	1977	1977	4C	D	B-E	1	Dragline only
320-B	8	1923	1930	8C R	S E	Bucyrus	37	Stripping shovel. Dragline: "Class 320"

Model	Standard Dipper c.y.	Manufactured From	To	Mounting C R W S*	Power S G D E A*	Original Design by	Number Built	Notes
375-B	7	1929	1940	8C	E	Bucyrus	7	Stripping shovel. Dragline: "Class 375"
385-B	12	1929	1929	8C	E	Bucyrus	1	Stripping shovel.
395-B	34	1979	1988	C	E	B-E	18	
395-B I	42	1992	1993	C	E	B-E	8	
395-B II	39	1989	1994	C	E	B-E	4	
395-B III	44	1995	1998	C	E	B-E	4	
495-B	53	1990	1996	C	E	B-E	13	
495-B I	55	1996	2000	C	E	B-E	25	
495-B II	55	2000	2002	C	E	BI	n/a	
495HD	60	2003	Current	C	E	BI	n/a	
495HR	60	2003	Current	C	E	BI	n/a	
495HF	60	2002	Current	C	E	BI	n/a	
550-B	20	1936	1954	8C	E	B-E	8	Stripping shovel.
595-B	57	1997	2001	C	E	Marion	4	Former Marion 351-M
750-B I	12	1927	1930	8C	E	Bucyrus	10	Stripping shovel. Non-counterbalanced hoist
750-B II	20	1930	1940	8C	E	B-E	4	Stripping shovel. Counterbalanced hoist
950-B	30-33	1935	1941	8C	E	B-E	11	Stripping shovel. Counterbalanced hoist
1050-B	33-45	1941	1960	8C	E	B-E	12	Stripping shovel. Counterbalanced hoist
1650-B	55-70	1956	1964	8C	E	B-E	5	Stripping shovel.
1850-B	90	1963	1963	8C	E	B-E	1	Stripping shovel.
1950-B	105-130	1965	1967	8C	E	B-E	2	Stripping shovel.
3850-B	115-140	1962	1964	8C	E	B-E	2	Stripping shovel.
					*	*		

*	*	*
C=Crawler	S=Steam	c = current machine
R=Rail	G=Gasoline	n/a = Information not available
W=Wheel	D=Diesel	
S=Skid	E=Electric	
T=Truck	A=Air	

APPENDIX 1

(B) WALKING DRAGLINES PRODUCTION AND SHIPPING DATES

Monighan / Bucyrus-Monighan / Bucyrus-Erie / Ruston-Bucyrus / Bucyrus International

Model	*Bucket Range c.y.*	*Manufactured From*	*To*	*No. Built*	*Original Design*
Martinson Tractor	1	1913	1913	1	Monighan
1-T	1	1913	1925	117	Monighan
1 ½ -T	1 ½ - 2	1922	1924	15	Monighan
2-T	2	1915	1922	26	Monighan
2 ½ T	2 ½	1922	1922	1	Monighan
3-T	3	1913	1925	65	Monighan
3 ½ T	3 ½	1922	1922	3	Monighan
4-T	4	1924	1925	6	Monighan
1-W	1	1926	1926	6	Monighan
1 ½ -W	1 ½	1926	1926	2	Monighan
2-W	2 - 2 ½	1926	1938	25	Monighan
3-W	3 - 3 ½	1925	1944	36	Monighan
3-W (R-B)	2 ½ - 3	1941	1955	15	Bucyrus-Monighan
4-W	4	1926	1932	8	Monighan
5-W	4 - 6	1934	1948	79	Bucyrus-Monighan
5-W (R-B)	3 - 5	1939	1971	62	Bucyrus-Monighan
6-W	5 - 6	1926	1934	24	Monighan
7-W	6 - 7	1942	1971	34	Bucyrus-Monighan
9-W	8 - 10	1938	1954	44	Bucyrus-Monighan
10-W	8 - 12	1934	1939	5	Bucyrus-Monighan
15-W	12 - 14	1940	1940	3	Bucyrus-Monighan
6150	6 - 8	1929	1932	9	Monighan
6160	6 - 8	1932	1938	16	Monighan
8160	8 - 10	1931	1931	3	Monighan
180-W	4 ½ - 6	1954	1965	16	Bucyrus-Erie
200-W	4 ½ - 6	1945	1956	55	Bucyrus-Monighan
380-W	10 - 16	1978	1985	18	Bucyrus-Erie
380-W (R-B)	10 - 12	1979	1985	6	Bucyrus-Erie
450-W	9 - 12	1948	1954	11	Bucyrus-Erie
480-W	13 - 18	1955	1979	35	Bucyrus-Erie
480-W (R-B)	15 - 18	1965	1978	10	Bucyrus-Erie
500-W	12 - 14	1946	1959	9	Bucyrus-Erie
650-B	15 - 17	1946	1954	13	Bucyrus-Erie
680-W	20 - 25	1982	1988	5	Bucyrus-Erie
770-B	19 - 21	1954	1965	9	Bucyrus-Erie
800-W	28	1966	1970	2	Bucyrus-Erie
950-B	12	1935	1935	1	Bucyrus-Erie
1150-B	20 - 25	1944	1950	17	Bucyrus-Erie
1250-B	25 - 38	1951	1958	8	Bucyrus-Erie
1250-W	35	1959	1963	6	Bucyrus-Erie
1260-W	30 - 40	1965	1990	27	Bucyrus-Erie
1260-W (R-B)	30 - 40	1976	1982	6	Bucyrus-Erie
1300-W	33 - 47	1971	1985	12	Bucyrus-Erie
1350-W	45 - 60	1967	1977	10	Bucyrus-Erie
1360-W	50	1975	1976	2	Bucyrus-Erie
1370-W	58 - 65	1970	1984	38	Bucyrus-Erie
1450-W	60	1963	1968	7	Bucyrus-Erie
1500-W	70	1970	1971	2	Bucyrus-Erie
1550-W	65	1968	1968	2	Bucyrus-Erie
1570-W	58 - 80	1973	1991	46	Bucyrus-Erie
2550-W	75	1964	1966	4	Bucyrus-Erie
2560-W	85 - 90	1969	1969	2	Bucyrus-Erie

Model	Bucket Range c.y.	Manufactured From	To	No. Built	Original Design
2570-W	100 - 115	1971	1983	27	Bucyrus-Erie
2570-WS	140 - 160	1990	2000	2	Bucyrus-Erie
3270-W	176	1977	1977	2	Bucyrus-Erie
4250-W	220	1968	1968	1	Bucyrus-Erie
W2000	31 - 44	1993	2001	6*	Ransomes & Rapier
8200	65 - 94	2003	2003	1**	Marion
8750 (AC or DC)	80 - 135	2005	2008	4***	Marion

* 6 shipped as Bucyrus machines, 15 previously shipped as R&R machines
** 1 shipped as Bucyrus machine, 34 previously shipped as Marion machines
*** 4 shipped as Bucyrus machines, 24 previousely shipped as Marion DC machines

APPENDIX 1

(C) HYDRAULIC EXCAVATORS
Bucyrus-Erie and Ruston-Bucyrus

Model	Standard Bucket c.y.	Manufactured From	To	Notes
H-2	1/4	1948	1952	Original Milwaukee Hydraulics model
H-3 I	3/8	1951	1960	Original Milwaukee Hydraulics model
H-3 II	3/8	1960	1981	
H-4	10-Ton	1967	1967	Special H-5
H-5 I	½	1955	1964	New design by B-E
H-5 II	½	1964	1981	
H-6	15-Ton	1967	1967	Special H-5
3-RB	½	1962	1965	U.K. Design by R-B
3-RB II	½	1966	1967	U.K. Design by R-B
15-H I	½	1966	1969	
15-H II	½	1969	1976	
15-RBH	½	1969	1974	Built in U.K. by R-B
20-H I	7/8	1965	1969	
20-H II	7/8	1969	1972	
20-H III	7/8	1972	1974	
20-H IV	7/8	1974	1978	
20-RBH	7/8	1967	1972	Built in U.K. by R-B
21-H	7/8	1974	1977	
30-H I	1 1/4	1967	1969	
30-H II	1 1/4	1969	1979	
30-RBH	1 1/4	1969	1974	Built in U.K. by R-B
31-H	1 3/4	1973	1977	
40-H	3	1970	1982	
40-H IPA	5	1974	1980	
65-HL	--	1974	1980	Special log loader
150-RH	3/4	1976	1980	U.K. Design by R-B
155-RH	3/4	1980	1983	U.K. Design by R-B
175-RH	1	1983	1983	150-RH upper, 220-RH lower
220-RH	1 1/4	1976	1983	U.K. Design by R-B
220-RS	1 1/2	1980	1983	220-RH with shovel front
300-H	1	1977	1983	
325-H	1 3/4	1977	1984	
350-H	2 1/2	1975	1983	
375-RS	3	1981	1983	U.K. Design by R-B
400-H	4 1/4	1979	1985	
500-H	5 3/4	1979	1983	
550-HS	10	1982	1983	500-H with Shovel front

APPENDIX 1

(D) DRILLS

Bucyrus-Erie and Bucyrus-Armstrong

Model	*Manufactured* *From*	*To*	*Number Built*	*Type*
1-W	1955	1961	179	Water well drill
20-W	1949	1981	1,649	Water well drill
21-W	1935	1949	900	Water well drill
22-W	1940	1984	3,492	Water well drill
24-W	1938	1943	137	Water well drill
29-W	1935	1941	57	Water well drill
33-W	1933	1939	243	Water well drill
38-W	1933	1941	1	Water well drill
10-R / 10-RH	1960	1981	216	Rotary water well drill
2400-R	1975	1984	44	Rotary water well drill
2450-R	1975	1984	Incl. above	Rotary water well drill
24-L	1939	1953	811	Oil well drill
28-L	1945	1971	555	Oil well drill
36-L	1939	1984	1,023	Oil well drill
60-L	1953	1972	504	Oil well drill
48-L	1945	1959	91	Oil well drill
JPM-3	1953	1960	24	Jet Piercer
JPM-4	1962	1967	18	Jet Piercer
22-P	1941	n/a	n/a	Prospecting drill
23-P	1937	1941	n/a	Prospecting drill
26-P	1933	1937	n/a	Prospecting drill
27-P	1939	1939	1	Prospecting drill
33-P	1934	1937	n/a	Truck version of 26-P
22-T	1942	1962	243	Blast hole drill, churn type
26-T	1933	1937	31	Blast hole drill, churn type
27-T	1937	1961	373	Blast hole drill, churn type
29-T	1933	1959	753	Blast hole drill, churn type
38-T	1936	1937	n/a	Blast hole drill, churn type
42-T	1937	1953	470	Blast hole drill, churn type
50-T	1953	1964	51	Blast hole drill, churn type
30-R	1959	1974	93	Blast hole drill, rotary type
35-R	1986	1992	15	Blast hole drill, rotary type
33-35-37R / HR	2004	Current	n/a	Blast hole drill, rotary type
39-R / HR	1996	Current	n/a	Blast hole drill, rotary type
40-R	1955	1984	281	Blast hole drill, rotary type
45-R	1966	1989	316	Blast hole drill, rotary type
47-R	1983	1991	13	Blast hole drill, rotary type
49-R / HR	1986	Current	n/a	Blast hole drill, rotary type
50-R	1952	1972	149	Blast hole drill, rotary type
55-R	1972	1985	23	Blast hole drill, rotary type
59-R / HR	1992	Current	n/a	Blast hole drill, rotary type
60-R	1962	1993	374	Blast hole drill, rotary type
61-R	1962	1987	Incl. above	Blast hole drill, rotary type
67-R	1986	1987	3	Blast hole drill, rotary type
Winkie GW4	1959	1961	163	Diamond core drill
120-SL	1959	1960	n/a	Well service drill
No. 8	1933	1960	196	Bit dresser
No. 12	1933	1962	225	Bit dresser

APPENDIX 2

BUCYRUS CABLE EXCAVATORS BY DESIGN GROUPS

In the design of Bucyrus and Bucyrus-Erie cable excavators, the design of certain successful new models was developed into further models of different sizes. This development resulted in families of machines of similar design whose ancestry can be traced to a particular model or era of manufacture. The groups in this appendix include examples of similarly-designed machines wit their slight differences noted.

GROUP A. [First machine of this design: 1030 (Erie design)]
- External swing rack with centre pintle
- Bevel gear swing drive
- Twin drums, single shaft
- Spur gear engine power take-off
- Twin stick, rack & pinion chain-drive crowd shovel
- Worm drive boom hoist from swing shaft

Model	***Introduced***	***Special Features***
1020	1929	
1025	1930	
1030	1928	
1035	1929	
2030	1929	Special railroad excavator.
21-B	1931	
24-B	1938	Lincoln-built 24-RB machine only.
29-B	1937	Ratchet digging lock
32-B	1931	
33-B	1937	
34-B	1933	
37-B	1931	
43-B	1931	
44-B	1939	
45-B	1932	Long boom crane/dragline/clam only.
46-B	1933	Long boom crane/dragline/clam only.
48-B	1936	Separate bevel gears for swing and propel. Propel driven from reversing crowd shaft.
52-B	1930	Separate bevel gears for swing and propel. Air controlled main clutches. Automatic rope crowd.
55-B	1935	Separate bevel gears for swing and propel. Propel driven from reversing crowd shaft. Air controlled main clutches.

GROUP B. [First machine of this design: 10-B]
- Internal swing rack with hook rollers
- Spur gear swing drive
- Tandem drums, separate shafts
- Roller chain engine power take-off
- Single stick rope crowd shovel
- Boom hoist clutch on front drum shaft

Model	*Introduced*	*Special Features*
10-B	1934	Friction steering
11-B	1958	Truck mounted. Ball bearing swing. Boom hoist on rear drum shaft.
14-B	1960	Ball bearing swing. Boom hoist on rear drum shaft. Air control for lower works.
15-B	1940	
17-B	1936	
18-B	1938	
19-B	1936	
20-B	1938	
22-B	1937	Air control after 1966
25-B	1961	Air control
30-B	1956	Air control. Boom hoist from engine power take-off. Ball bearing swing after 1981

<u>GROUP C.</u> [First machine of this design: 54-B]

- Internal swing rack with hook rollers
- Bevel gear swing drive
- Twin drums, single shaft
- Roller chain engine power take-off
- Single stick rope crowd shovel
- Boom hoist from engine power take-off

Model	*Introduced*	*Special Features*
38-B	1945	Air control after 1963
51-B	1946	Boom hoist from separate bevel gear set.
54-B	1939	Twin stick, chain crowd on shovel.
61-B	1963	Ball bearing swing. Air control. Dual crowd.
62-B	1975	Ball bearing swing. Air control.
71-B	1954	Dual crowd. Air control.
73-B	1964	Special crane/dragline/clam only.
84-B	1964	External swing rack with hook rollers. Worm drive boom hoist from swing shaft. Air control. Dual crowd.
88-B	1950	External swing rack with hook rollers. Worm drive boom hoist from swing shaft. Air control. Dual crowd.

<u>GROUP D.</u> Electric Machines [First machine of this design: 120-B]

- External swing rack with centre pintle
- Propel drive from hoist counter-shaft
- Single-piece boom includes crowd motor
- Twin stick, rack & pinion electric crowd shovel
- Ward-Leonard control

Model	*Introduced*	*Special Features*
42-B	1930	Propel drive from hoist bull gear.
75-B	1928	
85-B	1935	
100-B	1926	
120-B	1925	
170-B	1929	

GROUP E. Electric Machines [First machine of this design: 110-B]
- External swing rack with centre pintle
- Propel drive from hoist motor, separate gear train
- Two-piece pinned shovel boom
- Crowd machinery on deck
- Single tubular stick, rope crowd shovel
- Ward-Leonard control

Model	*Introduced*	*Special Features*
110-B	1950	
115-B	1962	Special crane/dragline/clam only.
150-B	1951	
155-B	1975	
190-B	1953	
195-B	1968	
195-B I	1980	Lower works propel
200-B	1970	Dragline only. Propel drive from drag motor.

GROUP F. Electric Machines [First machine of this design: 270-B]
- External swing rack with centre pintle
- Lower works propel
- Single-piece boom
- Crowd machinery on deck
- Single tubular stick, rope crowd shovel

Model	*Introduced*	*Special Features*
210-B	1961	Dragline only. Propel drive from drag motor
270-B	1960	Propel drive from drag motor.
280-B	1962	Propel drive from drag motor.
290-B	1976	
290-B II	1982	Choice of AC or DC controlled frequency system.
295-B	1972	
295-B II	1980	Choice of AC or DC controlled frequency system.
395-B	1979	AC controlled frequency system.
495-B	1990	AC controlled frequency system.

APPENDIX 3

MANUFACTURING FACILITIES

BUCYRUS INTERNATIONAL, INC. [AND FORMER BUCYRUS-ERIE COMPANY]

CITY	*STATE/COUNTRY*	*OPENED*	*SOLD/CLOSED*
Bucyrus	Ohio	1880	1893
South Milwaukee	Wisconsin	1893	Continues
Evansville*	Indiana	1912	1982
Erie	Pennsylvania	1927	1984
Lincoln (Ruston-Bucyrus) **	England	1930	1985
Chicago (Bucyrus-Monighan)	Illinois	1931	1958
Erie (National Erie Corp.)	Pennsylvania	1951	1961
Guelph, Ontario	Canada	1957	1971
Richmond	Indiana	1958	1959
West Pittston (Atlas Chain)	Pennsylvania	1968	1974
Cicero (Brad Foote Gear)	Illinois	1969	1987
Pittsburgh (Pittsburgh Gear)	Pennsylvania	1969	1987
Lake Bluff (Hy-Dynamic)	Illinois	1971	1973
Glassport	Pennsylvania	1974	1981
Pocatello	Idaho	1974	1983
Racine	Wisconsin	1974	1982
Milwaukee	Wisconsin	2005	Continues

* Evansville closed 1914 – 1915, 1932 – 1936.
** Joint venture Ruston-Bucyrus Ltd. with Ruston & Hornsby Ltd.

APPENDIX 4 – GLOSSARY

There aren't enough words in the English language to describe the activities, machines and components used in the heavy equipment industry. Consequently most words used have another meaning when used in a different context. Here are some of the common words used in the heavy equipment industry:

A-Frame:	A gantry consisting of a simple triangular frame resembling the letter 'A'.
Apron:	The moveable front gate on a scraper bowl.
Backhoe:	An excavator attachment consisting of boom, stick and dipper which is pulled toward the machine to collect its load. May also refer to the entire excavator.
Bail:	A hinged arch or yoke connecting a dipper to its hoisting sheaves.
Base Machine:	A fully-revolving excavator without attachments.
Boom:	The hinged projecting structure at the front of an excavator.
Bowl:	Bucket or box on a scraper in which the material is carried.
Car:	A rail-mounted truck or carriage.
Carbody:	The lower works on an excavator supporting crawler tracks or wheels. Derived from the original steam shovels which ran on railroad cars.
Crowd, automatic:	Crowd drive on a cable shovel that is inter-connected with the hoist.
Crowd, dual	Crowd drive on a cable shovel with both automatic and independent capabilities.
Crowd, independent:	Crowd drive on a cable shovel that is independent from the hoist.
Crowd:	The motion of a shovel which pushes the dipper arm into the bank.
Crown sheave:	The cable sheave at the top of a drill mast or tower.
Dipper:	The bucket of an excavator when equipped as a shovel or backhoe.
Draw Works:	The winch machinery on a crane or cable excavator.
Drilling string:	Assembly of drill bit, tools, drill stem, rope etc. connected to the drill hoist.
Ejector:	The moveable rear wall of a scraper bowl which ejects the load.
Excavator:	A machine used for excavating whose digging cycle is usually performed from one position without propelling. Includes shovel, backhoe, dragline, clamshell, skimmer, bucket wheel, and bucket chain types.
Front End:	The attachment on the front of an excavator which does the digging.
Gantry:	An upward extension above the house on an excavator to support the boom.
Grousers:	Cleats in crawler shoes.
Handle:	The arm or dipper stick on which the dipper is held on a cable shovel.
Hook Roller:	A roller and bearing fixed to the underside of the revolving frame of an excavator which engages a circular channel on the lower works.
House:	The structure on the revolving frame of an excavator covering the machinery and including the operator's cab.
Lower Works:	All machinery and structural components below the swing circle on a revolving excavator.
Mast:	A tower structure positioned between the boom and gantry on a crane or dragline to support

the boom. The vertical tower on a drill.

Padlock Sheave: Hinged sheave assembly attached to a shovel dipper.

Railroad Shovel: An excavator that swings only its front end during the digging cycle, usually steam powered. Can be mounted on rails or wheels.

Retract: The motion on a cable shovel that pulls the dipper away from the bank.

Sheave: Pulley or wheel with circumferential groove supporting a moveable cable.

Shipper shaft: The shaft located near the center of a shovel boom on which the dipper stick pivots or slides.

Shovel: Excavator attachment consisting of boom, stick and dipper which digs away from the machine to collect its load. May also refer to the entire excavator.

Skimmer: Excavator attachment on which the dipper runs along the boom to collect its load.

Spudder: A churn-type drill that drills holes by lifting and dropping a chisel bit.

Stick: The arm or handle attached to a shovel dipper.

Stripping Shovel: A revolving crawler-mounted excavator with a long boom capable of casting the excavated material into piles instead of loading a truck.

Struck Load: The volume of a “water level” load, i.e. without heaping.

Swing Circle: The anti-friction bearing between the upper and lower works on an excavator.

Swing: The action of rotating or slewing an excavator or crane about its centre point.

Transit Crane: A truck mounted mobile crane.

Tumbler: The driving wheel in a crawler assembly with indentations matching those in the crawler shoes.

Universal Excavator: An excavator capable of interchanging its front end.

Upper Works: All machinery and structural components above the swing circle on a revolving excavator.

Ward-Leonard: An electric drive system utilizing AC drive motors to generate DC power for precise variable speed of DC motors. Maximum torque is available at stall speed.

TRACTORS & CONSTRUCTION EQUIPMENT

Bucyrus Heavy Equipment: Construction and Mining Machines 1880-2008: A Photo Gallery ISBN 1-58388-219-7
Caterpillar Earthmovers at Work: A Photo Gallery ISBN 1-58388-120-4
Caterpillar Photo Gallery ISBN 1-882256-70-0
Caterpillar Pocket Guide The Track-Type Tractors 1925-1957 ISBN 1-58388-022-4
Caterpillar D-2 & R-2 Photo Archive ISBN 1-882256-99-9
Caterpillar D-8 1933-1974 Photo Archive Incl. Diesel 75 & RD-8 ISBN 1-882256-96-4
Caterpillar Military Tractors Volume 1 Photo Archive ISBN 1-882256-16-6
Caterpillar Military Tractors Volume 2 Photo Archive ISBN 1-882256-17-4
Caterpillar Sixty Photo Archive ISBN 1-882256-05-0
Caterpillar Ten Photo Archive Incl. 7c Fifteen & High Fifteen ISBN 1-58388-011-9
Circus & Carnival Tractors 1930-2001 Photo Archive ISBN 1-58388-076-3
Cletrac and Oliver Crawlers Photo Archive ISBN 1-882256-43-3
Classic American Steamrollers 1871-1935 Photo Archive ISBN 1-58388-038-0
Euclid Earth-Moving Equipment 1924-1968: A Photo Gallery ISBN 1-58388-129-8
Farmall F-Series Photo Archive ISBN 1-882256-02-6
Farmall Super Series Photo Archive ISBN 1-882256-49-2
Fordson 1917-1928 Photo Archive ISBN 1-882256-33-6
Hart-Parr Photo Archive ISBN 1-882256-08-5
Holt Tractors Photo Archive ISBN 1-882256-10-7
International TracTracTor Photo Archive ISBN 1-882256-48-4
John Deere Model A Photo Archive ISBN 1-882256-12-3
John Deere Model D Photo Archive ISBN 1-882256-00-X
Marion Construction Machinery 1884-1975 Photo Archive ISBN 1-58388-060-7
Marion Mining & Dredging Machines Photo Archive ISBN 1-58388-088-7
R. G. LeTourneau Heavy Equipment: The Mechanical Drive Era 1921-1953 A Photo Gallery ISBN 1-58388-214-6
R. G. LeTourneau Heavy Equipment: The Electric-Drive Era (1953-1971) ISBN 1-58388-226-X

TRUCKS

AM General: Hummers, Mutts, Buses & Postal Jeeps ISBN 1-58388-135-2
Autocar Trucks 1899-1950 Photo Archive ISBN 1-58388-115-8
*Brockway Trucks 1948-1961 Photo Archive** ISBN 1-882256-55-7
Chevrolet El Camino Photo History Incl. GMC Sprint & Caballero ISBN 1-58388-044-5
Circus and Carnival Trucks 1923-2000 Photo Archive ISBN 1-58388-048-8
Diamond T Trucks 1911-1966 Photo Archive ISBN 1-58388-204-9
Dodge C-Series Trucks Restorer's & Collector's Reference Guide and History ISBN 1-58388-140-9
Dodge Heavy-Duty Trucks 1928-1975 ISBN 1-58388-194-8
Dodge Pickups 1939-1978 Photo Album ISBN 1-882256-82-4
Dodge Ram Trucks 1994-2001 Photo History ISBN 1-58388-051-8
Dodge Trucks 1948-1960 Photo Archive ISBN 1-882256-37-9
El Camino by Chevrolet ISBN 1-58388-215-4
Federal Trucks Photo Archive ISBN 1-58388-223-5
Ford Heavy-Duty Trucks 1948-1998 Photo History ISBN 1-58388-043-7
Ford Medium-Duty Trucks 1917-1998 Photo History ISBN 1-58388-162-X
Ford Ranchero 1957-1979 Photo History ISBN 1-58388-126-3
Freightliner Trucks 1937-1981 Photo Archive ISBN 1-58388-090-9
FWD Trucks 1910-1974 Photo Archive ISBN 1-58388-142-5
GMC Heavy-Duty Trucks 1927-1987 ISBN 1-58388-125-5
GMC Light-Duty Trucks ISBN 1-58388-191-3
International Heavy Trucks of the 1950s: At Work ISBN 1-58388-160-3
International Heavy Trucks of the 1960s: At Work ISBN 1-58388-161-1
Jeep 1941-2000 Photo Archive ISBN 1-58388-021-6
Jeep Prototypes & Concept Vehicles Photo Archive ISBN 1-58388-033-X
Kenworth Trucks 1950-1979 At Work ISBN 1-58388-147-6
Land Rover: The Incomparable 4x4 from Series 1 to Defender, Ludvigsen Library Series ISBN 1-58388-179-4
*Mack Model AB Photo Archive** ISBN 1-882256-18-2
*Mack AP Super-Duty Trucks 1926-1938 Photo Archive** ISBN 1-882256-54-9
*Mack Model B 1953-1966 Volume 2 Photo Archive** ISBN 1-882256-34-4
*Mack EB-EC-ED-EE-EF-EG-DE 1936-1951 Photo Archive** ISBN 1-882256-29-8
*Mack FC-FCSW-NW 1936-1947 Photo Archive** ISBN 1-882256-28-X
*Mack LF-LH-LJ-LM-LT 1940-1956 Photo Archive** ISBN 1-882256-38-7
Muscle Trucks: High-Performance Pickups ISBN 1-58388-197-2
Peterbilt Trucks 1939-1979 At Work ISBN 1-58388-152-2
Refuse Trucks Photo Archive ISBN 1-58388-042-9
Reo Trucks 1910-1966 Photo Archive ISBN 1-58388-181-6
Semi Trucks of the 1950s: A Photo Gallery ISBN 1-58388-187-5
Studebaker Trucks 1927-1940 Photo Archive ISBN 1-882256-40-9
The Long Haul: American Trucking Companies ISBN 1-58388-211-1
White Trucks 1900-1937 Photo Archive ISBN 1-882256-80-8

BUSES

Buses of ACF Photo Archive Including ACF-Brill And CCF-Brill ISBN 1-58388-101-8
Buses of Motor Coach Industries 1932-2000 Photo Archive ISBN 1-58388-039-9
City Transit Buses of the 20th Century Photo Gallery ISBN 1-58388-146-8
Fageol & Twin Coach Buses 1922-1956 Photo Archive ISBN 1-58388-075-5
Flxible Intercity Buses 1924-1970 Photo Archive ISBN 1-58388-108-5
Flxible Transit Buses 1953-1995 Photo Archive ISBN 1-58388-053-4
GM Intercity Coaches 1944-1980 Photo Archive ISBN 1-58388-099-2
Greyhound in Postcards: Buses, Depots and Posthouses ISBN 1-58388-130-1
Highway Buses of the 20th Century Photo Gallery ISBN 1-58388-121-2
*Mack® Buses 1900-1960 Photo Archive** ISBN 1-58388-020-8
New York City Transit Buses 1945-1975 Photo Archive ISBN 1-58388-149-2
Prevost Buses 1924-2002 Photo Archive ISBN 1-58388-083-6
Rapid Transit Series Buses: General Motors and Beyond ISBN 1-58388-209-X
Trailways Buses 1936-2001 Photo Archive ISBN 1-58388-029-1
Trolley Buses 1913-2001 Photo Archive ISBN 1-58388-057-7
Trolley Buses Around the World: A Photo Gallery ISBN 1-58388-175-1
Welcome Aboard the GM New Look Bus: An Enthusiast's Reference ISBN 1-58388-167-5
Yellow Coach Buses 1923-1943 Photo Archive ISBN 1-58388-054-2

EMERGENCY VEHICLES

100 Years of American LaFrance: An Illustrated History ISBN 1-58388-139-5
Ahrens-Fox Fire Apparatus Photo Archive ISBN 1-58388-217-0
American Fire Apparatus Co. 1922-1993 Photo Archive ISBN 1-58388-131-X
American LaFrance 400 Series Photo Archive ISBN 1-58388-185-9
American LaFrance 500/600 Series: Photo Archive ISBN 1-58388-210-3
American LaFrance 700 Series 1945-1952 Photo Archive ISBN 1-882256-90-5
American LaFrance 700 Series 1945-1952 Photo Archive Volume 2 ISBN 1-58388-025-9
American LaFrance 700 & 800 Series 1953-1958 Photo Archive ISBN 1-882256-91-3
American LaFrance 900 Series 1958-1964 Photo Archive ISBN 1-58388-002-X
Classic Seagrave 1935-1951 Photo Archive ISBN 1-58388-034-8
Crown Firecoach 1951-1985 Photo Archive ISBN 1-58388-047-X
Emergency One Fire Apparatus 1974-2007 ISBN 1-58388-225-1
Elevating Platforms: A Fire Apparatus Photo Gallery ISBN 1-58388-164-6
Encyclopedia of Canadian Fire Apparatus ISBN 1-58388-119-0
Fire Chief Cars 1900-1997 Photo Album ISBN 1-882256-87-5
Firefighting Tanker Trucks and Tenders: A Fire Apparatus Photo Gallery ISBN 1-58388-138-7
Funeral Service & Flower Cars Photo Archive ISBN 1-58388-227-8
FWD Fire Trucks 1914-1963 Photo Archive ISBN 1-58388-156-5
Grumman Fire Apparatus 1976-1992 Photo Archive ISBN 1-58388-165-4
Hahn Fire Apparatus 1923-1990 Photo Archive ISBN 1-58388-077-1
Heavy Rescue Trucks 1931-2000 Photo Gallery ISBN 1-58388-045-3
Imperial Fire Apparatus 1969-1976 Photo Archive ISBN 1-58388-091-7
Industrial and Private Fire Apparatus 1925-2001 Photo Archive ISBN 1-58388-049-6
*Mack Fire Trucks 1911-2005 Illustrated History** ISBN 1-58388-157-3
Maxim Fire Apparatus 1914-1989 Photo Archive ISBN 1-58388-050-X
Maxim Fire Apparatus Photo History ISBN 1-58388-111-5
Navy & Marine Corps Fire Apparatus 1836 -2000 Photo Gallery ISBN 1-58388-031-3
Pierre Thibault Ltd. Fire Apparatus 1918-1990 Photo Archive ISBN 1-58388-074-7
Pierce Fire Apparatus 1939-2006 Illustrated History ISBN 1-58388-189-1
Pirsch Fire Apparatus 1890-1991 Photo Archive ISBN 1-58388-082-8
Police Cars: Restoring, Collecting & Showing America's Finest Sedans ISBN 1-58388-046-1
Saulsbury Fire Rescue Apparatus 1956-2003 Photo Archive ISBN 1-58388-106-9
Seagrave 70th Anniversary Series Photo Archive ISBN 1-58388-001-1
Seagrave Fire Apparatus 1959-2004 Photo Archive ISBN 1-58388-132-8
TASC Fire Apparatus 1946-1985 Photo Archive ISBN 1-58388-065-8
The Illustrated Encyclopedia of American Fire Engine Manufacturers ISBN 1-58388-203-0
Van Pelt Fire Apparatus 1925-1987 Photo Archive ISBN 1-58388-143-3
Volunteer & Rural Fire Apparatus Photo Gallery ISBN 1-58388-005-4
W.S. Darley & Co. Fire Apparatus 1908-2000 Photo Archive ISBN 1-58388-061-5
Wildland Fire Apparatus 1940-2001 Photo Gallery ISBN 1-58388-056-9
Young Fire Equipment 1932-1991 Photo Archive ISBN 1-58388-015-1

More Great Titles From
Iconografix

All Iconografix books are available from direct mail specialty book dealers and bookstores worldwide, or can be ordered from the publisher. For book trade and distribution information or to add your name to our mailing list and receive a **FREE CATALOG** contact:

Iconografix, Inc.
PO Box 446, Dept BK
Hudson, WI, 54016

Telephone: (715) 381-9755,
(800) 289-3504 (USA),
Fax: (715) 381-9756
info@iconografixinc.com
www.iconografixinc.com

More great books from Iconografix

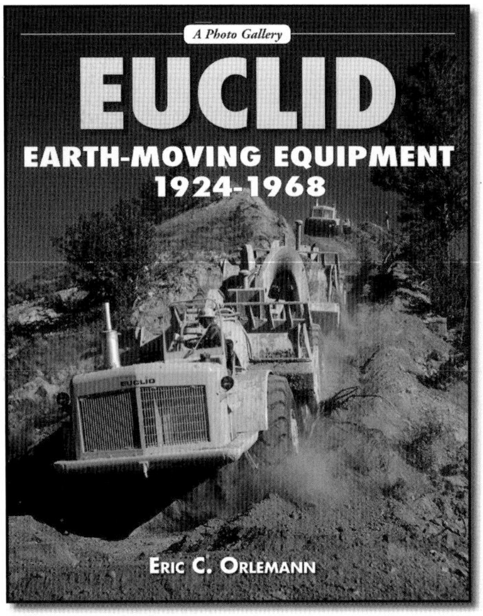

ISBN 1-58388-129-8

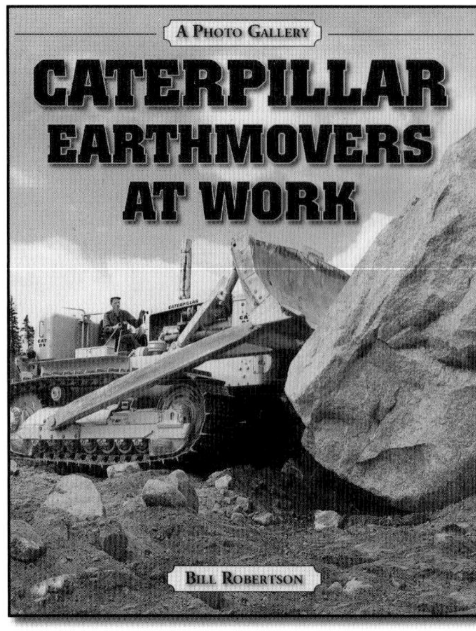

ISBN 1-58388-120-4

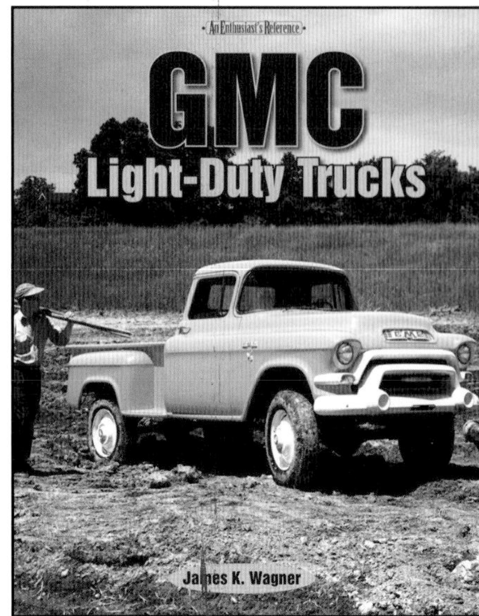

ISBN 978-1-58388-191-0

ISBN 978-1-58388-187-3

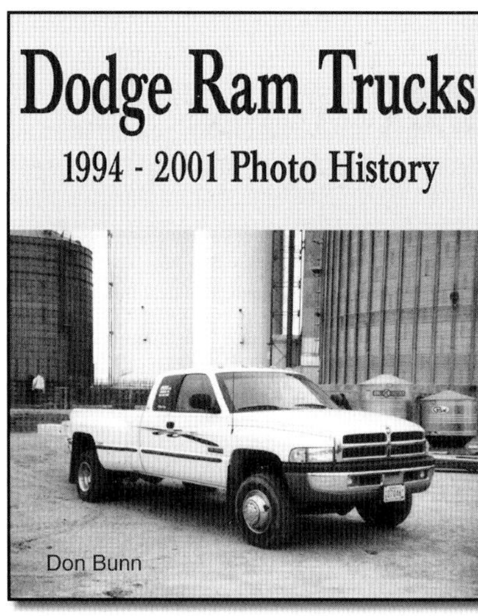

ISBN 1-58388-051-8

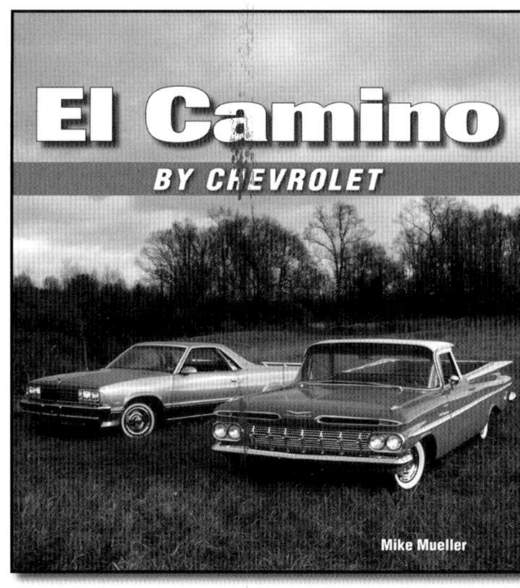

ISBN 978-1-58388-215-3

ISBN 1-58388-038-0

Iconografix, Inc.
P.O. Box 446, Dept BK,
Hudson, WI 54016
For a free catalog call: 1-800-289-3504
info@iconografixinc.com
www.iconografixinc.com